Techniques in

GENETIC ENGINEERING

Techniques in GENETIC ENGINEERING

Işıl Aksan Kurnaz

CRC Press
Taylor & Francis Group
Boca Raton London New York

CRC Press is an imprint of the
Taylor & Francis Group, an informa business

CRC Press
Taylor & Francis Group
6000 Broken Sound Parkway NW, Suite 300
Boca Raton, FL 33487-2742

Printed on acid-free paper
Version Date: 20150409

International Standard Book Number-13: 978-1-4822-6089-2 (Hardback)

Visit the Taylor & Francis Web site at
http://www.taylorandfrancis.com

and the CRC Press Web site at
http://www.crcpress.com

To my husband and partner in life,

my two adorable kids,

my parents and my family,

my devoted assistants,

and all my students …

We must not forget that when radium was discovered no one knew that it would prove useful in hospitals. The work was done of pure science. And this is a proof that scientific work must not be considered from the point of view of the direct usefulness of it. It must be done for itself, for the beauty of science, and then there is always the chance that a scientific discovery may become like the radium a benefit for humanity.

Marie Curie (1867–1934)
Lecture at Vassar College, May 14, 1921

Contents

Preface

This undergraduate textbook is a collection of more than eight years of lecture materials that I had prepared for the course *GBE 318 Techniques in Genetic Engineering*, and more recently, *GBE 341 Techniques in Genetic Engineering I* as well as *GBE 342 Techniques in Genetic Engineering II*, at Yeditepe University (Istanbul, Turkey). It must be noted that this book is not for the advanced audience, nor is it intended as a laboratory manual. In this textbook, I have tried to not only provide an up-to-date theoretical background for the student, but also to provide real-life case study problems and sample solutions, as these key elements are missing from many textbooks on this topic. Unfortunately, the book cannot cover all the assays and procedures used in genetic engineering. I could only incorporate some of the more common techniques into this book due to space constraints, but this information combined with the case studies are believed to be a good starting point for undergraduate students or newcomers to the field.

This book does not cover basic molecular biology, biochemistry of macromolecules, and other concepts—we assume that the reader already has a basic theoretical background in cell biology, molecular biology, and genetics. The reader is kindly directed to the "bibles" of the field, such as *Molecular Cloning*.* Instead, this book focuses on how to work with the

* Green, M. and Sambrook, J. (2012). *Molecular Cloning—A Laboratory Manual*, 4th Ed., Long Island, NY: CSHL Press.

genetic material or other biological macromolecules so as to engineer new combinations or products in the laboratory. As such, it should be viewed as an intermediate-level book that covers part of the applied genetics and molecular biology technologies for undergraduate studies or problem sessions. Therefore, molecular biology and genetics students, medical geneticists or biochemists, and clinicians with an interest in molecular biology among many others can benefit from this book in their more advanced studies.

Lecture Slides are available on the CRC Web site at: http://www.crcpress.com/product/isbn/9781482260892.

Acknowledgments

I would like to acknowledge Yeditepe University, our Faculty of Engineering and Architecture, and our Department of Genetics and Bioengineering, where I have offered *GBE 318*, *GBE 341*, and *GBE 342 Techniques in Genetic Engineering* courses over the years, which initiated the idea for writing this book. Very special thanks to all my students of *GBE 318*, *GBE 341*, and *GBE 342*, who have suffered enormously during the teaching of these courses over the years! I also need to acknowledge Ferruh Ozcan and Nagehan Ersoy Tunali for their encouragement during the writing of this book, and the anonymous reviewers for their help with the improvement of the quality.

More importantly, I would like to acknowledge, in alphabetical order: Göksu Alpay, Oya Arı, Başak Arslan, Uğur Dağ, Özlem Demir, Burcu Erdoğan, Başak Kandemir, Elif Kon, Eray Şahin, Melis Savaşan, and Perihan Ünver who have been right-hand men and women as my teaching and/or laboratory assistants during the courses.

> Most people say that it is the intellect which makes a great scientist. They are wrong: it is character.
>
> **Albert Einstein**

Abbreviations and Acronyms

A: Adenine (also used to designate adenosine triphosphate in nucleic acids)
A.D.: (1) Anno Domini; designates years after the birth of Christ
(2) Activation domain
ADA: Adenosine deaminase
AP: Alkaline phosphatase
APS: Ammonium persulfate
ARS: (1) Autonomous replicating sequence
(2) Aminoacyl-tRNA synthetase
ASC: Adult stem cell
asRNA: Antisense RNA
ATP: Adenosine (nucleoside) triphosphate
BAC: Bacterial artificial chromosome
B.C.: Before Christ: designates years before the birth of Christ
BCIP: 5-bromo-4-chloro-3-indolyl phosphate
BFP: Blue fluorescent protein
BLAST: Basic Local Alignment Search Tool
bp: Base pair
Bq: Becquerel
Bt: *Bacillus thuringiensis*
C: (1) Cytosine (also used to designate cytidine triphosphate in nucleic acids)
(2) Carbon

CaMV: Cauliflower mosaic virus
cDNA: Complementary DNA
CEN: Centromere
Ci: Curie
CMV: Cytomegalovirus
CRISPR: Clustered regularly interspaced short palindromic repeats
CTP: Cytidine triphosphate
Cy3: Cyanine 3
Cy5: Cyanine 5
Da: Dalton
dATP: Deoxyadenosine triphosphate
DBD: DNA-binding domain
dCTP: Deoxycytidine triphosphate
ddNTP: Dideoxynucleotide
dGTP: Deoxyguanosine triphosphate
DIG: Digoxigenin
DNA: Deoxyribonucleic acid
dNTP: Deoxyribonucleoside triphosphate (also called deoxynucleotide)
ds: Double stranded
DTT: Dithiothreitol
dTTP: Deoxythymidine triphosphate
EBI: European Bioinformatics Institute
EC: Embryonic carcinoma cell
EG: Embryonic germ cell
EGFP: Enhanced GFP
ELISA: Enzyme-linked immunosorbent assay
EMBL: European Molecular Biology Laboratory
EMSA: Electrophoretic Mobility Shift Assay
env: Envelope
ES: Embryonic stem cell
FDA: Food and Drug Administration
FFPE: Formalin fixed and paraffin embedded
FITC: Fluoroscein isothiocyanate
FP: Fluorescent protein

FRET: Fluorescence resonance energy transfer
G: Guanine (also used to designate guanosine triphosphate in nucleic acids)
G3PDH: Glyceraldehyde-3-phosphate dehydrogenase
gag: Group-specific antigen
GAPDH: Glyceraldehyde-3-phosphate dehydrogenase
GFP: Green fluorescent protein
GM: Genetically modified
GMO: Genetically modified organism
GST: Glutathione-S-transferase
GTP: Guanosine triphosphate
Gy: Gray
H: Hydrogen
HA: Hemagglutunin
hESC: Human embryonic stem cell
His: Histidine
HRP: Horseradish peroxidase
HSC: Hematopoietic stem cells
HUGO: Human Genome Organization
ICM: Inner cell mass
IF: Immunofluorescence
iGEM: International Genetically Engineered Machine
IgG: Immunoglobulin G
IHC: Immunohistochemistry
IP: Immunoprecipitation
iPSC: Induced pluripotent stem cell
IPTG: Isopropyl β-D-1-thiogalactopyranoside
IRES: Internal ribosome entry site
IVF: *In vitro* fertilization
J: Joule
kb: Kilobase
kDa: Kilodalton
LTR: Long terminal repeats
MALDI: Matrix-assisted laser desorption/ionization
MCS: Multiple cloning site
miRNA: MicroRNA

mRNA: Messenger RNA
MS: Mass spectrometry
NADH: Nicotinamide adenine dinucleotide (reduced form)
NBT: Nitroblue tetrazolium
NCBI: National Center for Biotechnology Information
NLS: Nuclear localization sequence
NTP: Nucleoside triphosphate (also called *nucleotide* or *ribonucleotide*)
-OH: Hydroxyl group
Ori: Origin of replication
P: Phosphorus
PCR: Polymerase chain reaction
PVDF: Polyvinylidene fluoride
qPCR: Quantitative (real-time) PCR
Q-RT-PCR: Quantitative (real-time) RT-PCR
RFLP: Restriction fragment length polymorphism
RNA: Ribonucleic acid
RNAi: RNA interference
RPA: Ribonuclease protection assay
RT: Reverse transcriptase; reverse transcription; (sometimes, real time)
RT-PCR: Reverse transcription polymerase chain reaction
S: Sulphur
SCID: Severe Combined Immunodeficiency syndrome
SCNT: Somatic cell nuclear transfer
SDS: Sodium dodecyl sulfate
SDSC: San Diego Supercomputer Center
SDS-PAGE: SDS polyacrylamide gel electrophoresis
shRNA: Short hairpin RNA
siRNA: Short interfering RNA
ss: Single stranded
Sv: Sievert
SV40: Simian virus 40
T: Thymine (also used to designate thymidine triphosphate in nucleic acids)
TALEN: Transcription activator-like effector nucleases

TEL: Telomere
TEMED: Tetramethylethylenediamine
tet: Tetracycline
TetR: Tetracycline repressor
TGF-β: Transforming growth factor-β
Ti: Tumor inducing
TIGR: The Institute for Genetic Research
tRNA: Transfer RNA
U: Uracil (also used to designate uridine triphosphate in nucleic acids)
UTP: Uridine triphosphate
X-gal: 5-bromo-4-indoyl-B-D-galactopyranoside
Y1H: Yeast one-hybrid
Y2H: Yeast two-hybrid
Y3H: Yeast three-hybrid
YAC: Yeast artificial chromosome
YFP: Yellow fluorescent protein
ZFN: Zinc finger nucleases

Chapter 1

Introduction to Genetic Engineering

> Science may set limits to knowledge, but should not set limits to imagination.
>
> **Bertrand Russell (1872–1970)**

Man has used artificial selection to exploit and manipulate organisms for thousands of years—between 8000 and 1000 B.C. horses, camels, oxen, and many other species were already domesticated; by 6000 B.C. yeast was used to make beer; by 5000 B.C. plants such as maize, wheat, and rice were bred. Generation of life and reproduction has always been one of the major points of interest for ancient philosophers. One can almost imagine them sitting under a tree, observing nature around them, and trying to understand this mystical process. In 420 B.C. Socrates speculated on why children do not necessarily resemble their parents; by 400 B.C. Hippocrates would propose that males contribute to a child's character through semen: the idea of **heredity** was thus established.

It was not just Greeks or Romans who were in a constant quest for an answer to how life originates. Between A.D. 100–300 Hindu philosophers were giving much thought to the same questions of reproduction and inheritance. By the first millennium, they had already established the foundations of **genetics**; they observed that certain diseases might run in the family. They also came to believe, almost correctly, that children inherit all the characteristics of their parents. The Hindu laws stated, "A man of base descent can never escape his origin."

We have come a long way from Antonie van Leeuwenhoek's **homunculi**, the minute, preformed human beings that he believed to reside within the sperm in his micrographs. With the exponential increase in the number of biochemical studies during the 19th century, such as those on nucleic acids and amino acids, and the speeding up of the fermentation industry, biology took on a whole new direction. In 1864, Mendel presented his work on peas in a modest communication to the local Brunn Natural History Society, and published the results in 1865 in *Versuche uber Pflanzen Hybriden*. The work was largely neglected for quite some time, and the term **gene** or **genetics** was not yet coined.

Yet, in 1883 a new movement, **eugenics**, was being established under the leadership of Francis Galton in England, where genetic knowledge would be directly applied for the improvement of human existence. Eugenics was a movement most prominent in England, the United States, Germany, and several Scandinavian countries, and to a lesser extent in France and Russia, which lasted from the early 1900s to the mid-to-late 1930s. The movement attempted to use the recent revelations of Mendelian genetics to explain and resolve many social problems such as chronic unemployment and poverty, feeblemindedness, alcoholism, prostitution, rebelliousness, and criminality. By 1907, starting with Indiana in the United States, over half the states would pass state and federal laws that required sterilization of those considered "genetically inferior."

The next century saw a huge accumulation of data and know-how that would eventually lead to the first biotechnology products, including the use of agar described in 1882 by the Koch lab, the development of the autoclave in 1884 by a French company (Chamberland's Autoclaves), the discovery of X-rays by W. Roentgen in 1895, followed by the application of this information to X-ray crystallography by physicist Sir William Henry Bragg and his son William Lawrence Bragg and many others in 1913. However, most of the leap in technology in genetic engineering or recombinant DNA owes its progress to the physicists who became deeply interested in the biology of the cell after World War II.

Max Delbruck was a theoretical physicist—turned bacterial virologist. In 1949, he would write in *A Physicist Looks at Biology* (*Trans. Conn. Acad.* 38, p. 190):

> Biology is a very interesting field to enter for anyone, by the vastness of its structure and the extraordinary variety of strange facts it has collected. ... In biology we are not yet at the point where we are presented with clear paradoxes and this will not happen until the analysis of the behavior of living cells has been carried into far greater detail. This analysis should be done on the living cell's own terms...

Delbruck had invited his bacteriophage collaborator, Salvador Luria (1912–1991), to join him at Cold Spring Harbor Laboratory during the 1940s. Their combined research aim was to identify the physical nature of the gene. In 1943 Delbruck invited Alfred D. Hershey (1908–1997), who was then at Vanderbilt University working with bacteriophages, to come and work in his lab. In 1951, Hershey performed his famous "blender experiment" with his assistant Martha Chase, showing that the hereditary material is DNA and not protein. Luria and Hershey also demonstrated that bacteriophages mutated, and introduced criteria for distinguishing mutations from other modifications.

In 1945, William Astbury, a leading biophysicist in the field of X-ray diffraction analysis of structures of biological macromolecules, devised the term **molecular biology**. In the early

1950s, Rosalind Franklin and Maurice Wilkins obtained the X-ray diffraction data for DNA, which would prove crucial for Watson and Crick to establish their model of two helically intertwined chains tied together by hydrogen bonds between the purines and pyrimidines in 1953. At around the same time, bacterial plasmids were defined as autonomously replicating material, and in the late 1960s, Werner Arber identified the **restriction enzymes** in bacteria that were designed to cleave DNA. And in 1970, Temin and Baltimore independently identified the viral enzyme **reverse transcriptase**, which would result in the birth of recombinant DNA technology—the first recombinant DNA was produced in Boyer Laboratory in 1972, and in 1976, the first biotechnology company **Genentech** was born.

The big biotech boom would be seen in the 1980s, especially after the invention of the **polymerase chain reaction (PCR)** by Karen Mullis in 1983. Genentech's recombinant interferon gamma and Eli Lilly's recombinant human insulin appeared on the market in 1982. The Human Genome Initiative, later to be renamed the **Human Genome Project**, was launched in 1986 and its completion was announced nearly two decades later. Another biotech company, GenPharm International, Inc., created the first transgenic dairy cow to produce human milk proteins for infant formula in the 1990s, and in the same period the first authorized **gene therapy** began on a four-year-old girl with an immune disorder known as ADA, or adenosine deaminase deficiency. This hype was perhaps at its peak in 1997, when flash news came from Scotland's Roslin Institute that the first mammalian clone, Dolly the sheep, was born, through a procedure known as somatic cell nuclear transfer.

Now we have the complete genomes of many species, from bacteria to men; we have techniques to screen for genetic polymorphisms in individuals (such as those done for James Watson and Craig Venter), we can manipulate stem cells and generate knockout animals, transgenic animals, or even clones. We cannot foresee how much further the boundaries of genetic engineering can be pushed. This book merely seeks to give some basic background on the recent techniques employed in genetic engineering, and present the reader with real-life applications. We sincerely hope that many of the readers of this book will contribute to the expansion of the boundaries of molecular biology.

> In science the credit goes to the man who convinces the world, not the man to whom the idea first occurs.
>
> **Sir Francis Darwin (1848–1925)**
> *Eugenics Review, April 1914*

Chapter 2

Tools of Genetic Engineering

> Science is facts; just as houses are made of stones, so is science made of facts; but a pile of stones is not a house and a collection of facts is not necessarily science.
>
> **Henri Poincaré (1854–1912)**

The basic methodology for manipulating genes and cloning recombinant DNA molecules requires various restricting and modifying enzymes, certain methods for amplifying DNA, and vectors for carrying this recombinant DNA molecule (Hartl et al. 1988; Howe 2007; Nair 2008; Reece 2004). This chapter will introduce some of the basic tools required for cloning and genetic engineering. The following chapters will walk us through the more advanced tools for modifying the sequence of DNA or protein molecules and monitoring the effects of these changes.

This chapter will begin by explaining the history, function, and use of the restriction/modification system, then it will cover various vector systems including plasmids, cosmids, phage vectors, and specialist vectors, followed by some brief information on ligases for the joining of different DNA molecules, and finally it will cover the basics of cloning, bacterial transformation, and screening, as well as commonly used DNA techniques. After examining these basic tools and principles, the chapter will conclude with a problem session, where these principles will be applied to various examples.

2.1 Restriction Endonucleases

Nature does nothing uselessly.

Aristotle (384–322 B.C.)

In the 1960s, phage biologists worked on the mechanisms of host restriction, where bacteriophages isolated from one strain of *Eschericia coli* would infect bacteria of the same strain, but not other *E. coli* strains (in other words, the phage would be *restricted* in others). These studies led to the discovery and purification of the first restriction endonuclease, or restriction enzyme, from *E. coli* (Meselson and Yuan 1968). These restriction enzymes were observed to cut DNA into smaller fragments, and it was reasoned that since they appeared to be strain dependent, they would recognize and cut specific target sequences. Two years later, Smith and friends isolated another restriction enzyme from *Haemophilus influenza* strain Rd, called *HindII*, and showed that it recognizes a specific sequence of 5′-GT(T/C).(G/A)AC-3′, cleaving the sequence on both strands right in the middle (where the dot is) (Kelly and Smith 1970; Smith and Wilcox 1970). This has been accepted as the standard for abbreviation: the first enzyme that was isolated from *Serratia marcescens*, for instance, is known as *SmaI* and the third enzyme isolated from *Hemophilus influenzae*, strain d, is known as *HindIII*.

Bacteriophages invade bacteria as their host cell, and for any invader a defense mechanism will be developed. For bacteria, this mechanism is the host restriction/modification system, composed not only of the restriction endonuclease but also a methylating enzyme. Restriction endonucleases *cut* or *cleave* target sequences on phage DNA as part of this defense mechanism, whereby the phage will be rendered harmless. DNA of the invading phage will be digested at the recognition sites; however, the host DNA itself should be somehow *protected* from this digestion, which is achieved by modifying the bacterial DNA by addition of a methyl group to target recognition sites (modification). In other words, for an *E. coli* EcoRI restriction enzyme that recognizes the sequence 5′-GAATTC-3′ on phage DNA, there has to be an EcoRI methylase that modifies the same sequence in the bacterial genome (Figure 2.1).

There are four known different subtypes of restriction endonucleases that are grouped together based on the level of enzyme complexity, cofactor requirements, recognition sequence properties, and many other parameters.

2.1.1 Type I Endonucleases

These were the systems that were first characterized, and they consist of one enzyme with different subunits for recognition, cleavage, and methylation (all-in-one). Their mechanism of cleavage relies on the translocation of DNA until a mechanistic collision occurs (usually quite some distance away, up to 1000 bp), producing fairly random fragments. Thus, their lack of a specific cleavage point makes them unsuitable for specific gene cloning purposes.

EXAMPLE 2.1

EcoDR2 TCANNNNNN GTCG

2.1.2 Type II Endonucleases

These are the most common types of restriction enzymes, and most of the commercially available enzymes routinely

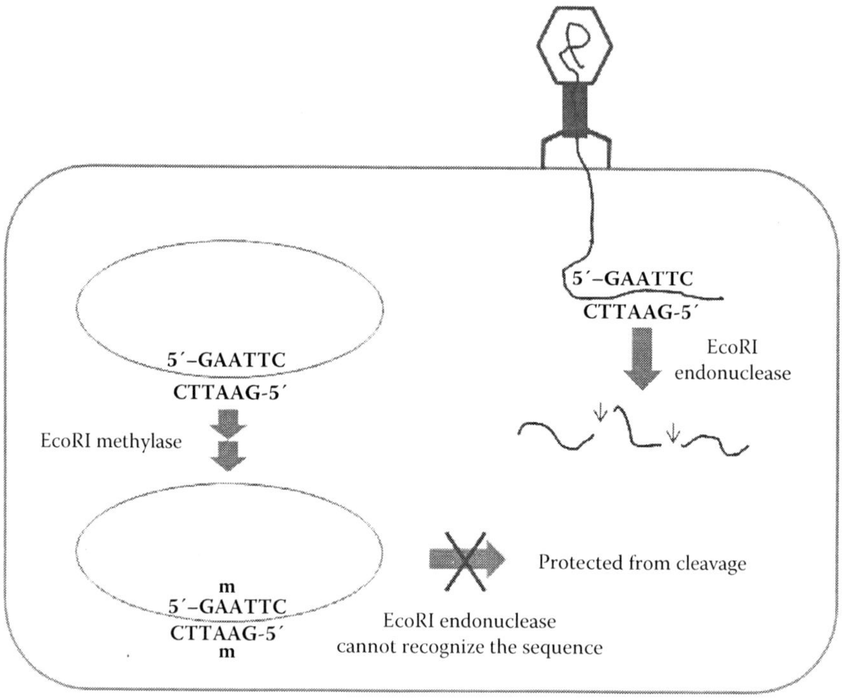

Figure 2.1 A schematic summary of the restriction/modification system. In this example, the recognition motifs for the EcoRI restriction endonuclease in the host genome are modified by the EcoRI methylase, which covalently adds a methyl group to the adenine nucleotide. This modification does not affect the structure of the host DNA, but simply disables the endonuclease from recognizing the motif, thus the host genome is protected from cleavage. Phage DNA, on the other hand, has not been previously methylated, therefore the EcoRI enzyme recognizes the cleavage site upon the entry of the phage DNA and cleaves it.

used in general cloning experiments fall into this category. There are two different genes for restriction and modification. They usually bind DNA as homodimers, thus recognizing symmetric sequences; however, some Type II enzymes can bind as heterodimers and recognize asymmetric sites—but no matter what, they are very specific and have fairly constant cut positions. The sites can either be *continuous*, as in the case of most 6-base-cutters (such as GAATTC for EcoRI),

Figure 2.2 The palindromic recognition motif for the EcoRI enzyme. The sequence reads the same in each orientation, and the nucleotides on either side of the midaxis are complementary.

or *discontinuous*, where the two half-sites are separated by a number of random nucleotides (such as GCCNNNNNGGC for BglI).

The most common Type II enzymes which are used in the laboratory recognize four to eight bases; but the recognition motif is palindromic, which means that not only does each strand read the same sequence 5′ to 3′, but the nucleotides are also *mirror images* through an axis that pass in the middle of the sequence (Figure 2.2).

Restriction enzymes, once they recognize their target sequence, may cut at different positions along the sequence (Figure 2.3). Restriction enzymes can either cut one or two nucleotides after the 5′ ends, as in the case of BamHI, right in the middle of the sequence, as in the case of EcoRV, or

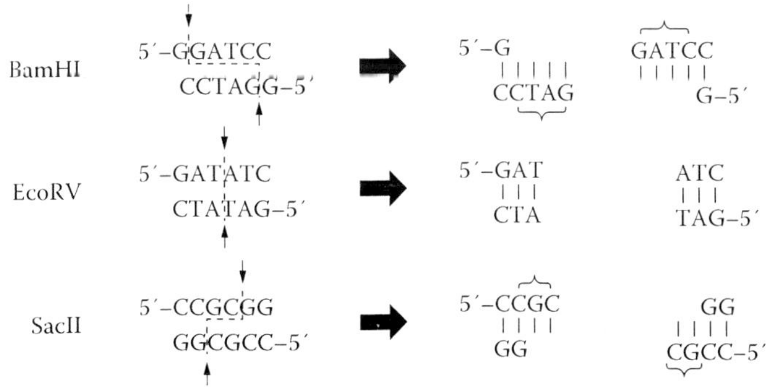

Figure 2.3 Different cut positions of restriction enzymes. BamHI cuts toward the 5′ end and generates a sticky end with a 3′ overhang, EcoRV cuts right in the middle of the sequence and generates a blunt end, and SacII cuts toward the 3′ end to generate a sticky end with a 5′ overhang.

four or five nucleotides after the 5′ end, as in the case of SacII in the examples above. Enzymes of the second group give rise to two separate double helices, where all nucleotides are paired (called the *blunt end*), whereas enzymes of the first and third group give rise to partially single-stranded ends that project out (called the *sticky ends* since they can "stick" another single-stranded region with a complementary sequence). Depending on whether the cut position is toward the 5′ end, as in the case of BamHI, or toward the 3′ end, as in the case of SacII, these sticky ends are called *3′ overhangs* or *5′ overhangs*, respectively.

2.1.3 Type IIs Endonucleases

This group of enzymes actually works in pairs, and recognizes asymmetric sequences. They cleave up to 20 bp away on one side of the recognition sequence. They are much more active on DNA containing multiple motifs. Their recognition site cannot be destroyed by blunting the ends of digestion products, which can be exploited in some cloning applications such as generating deletions along a DNA molecule.

2.1.4 Type III Endonucleases

Enzymes of Type III consist of one complex of two subunits encoded by the *mod* (modification) and *res* (restriction) genes. The recognition sequence is a set of two copies of nonpalindromic sites in inverse orientation. The enzyme then cleaves at a specific distance (24 to 26 bp) away from one of the copies. Since the exact cut site is not predetermined, these enzymes are nonsuitable for cloning purposes.

2.1.5 Type IV Endonucleases

These enzymes can recognize modified (even methylated) DNA. They are rather large proteins with two catalytic subunits, and they cleave outside their recognition sites. They can be subgrouped among themselves: enzymes that recognize *continuous* sequences (such as CTGAAG for Eco57I) cleave on one side only, whereas those that recognize *discontinuous* sequences (such as CGANNNNNNTGC for BcgI) cleave on both sides of the motif.

2.1.6 Isoschizomers and Neoschizomers

There are many different phages infecting a variety of different host strains, all of which have developed restriction/modification systems as a form of self-defense. Therefore, it is not surprising that enzymes from different bacteria may recognize the same recognition motif if they are infected with phage carrying the very same motif. If the two enzymes isolated from different bacteria (hence different enzyme names) recognize the same sequence and cut at exactly the same position, these enzymes are called *isoschizomers*; if the two enzymes recognize the same DNA sequence motif but cleave at different positions, then they are called *neoschizomers* (Figure 2.4) (Table 2.1).

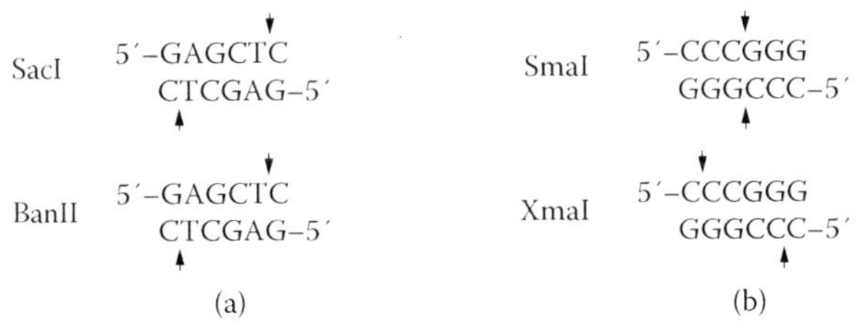

Figure 2.4 Isoschizomers and neoschizomers. (a) SacI and BanII in this example recognize exactly the same recognition motif and cut from exactly the same position, hence they are called *isoschizomers* of each other. (b) SmaI and XmaI in this example recognize the same motif but cut at different positions, hence they are neoschizomers. See Table 2.1 for more examples.

Table 2.1 Isoschizomers of Some Enzymes and Their Recognition Motifs as Examples

Enzyme	*Isoschizomer(s)*	*Recognition Sequence*
AclNI	SpeI	A/CTAGT
Bsp19I	NcoI	C/CATGG
Bsp106I	ClaI, Bsu15I	AT/CGAT
Eco32I	EcoRV	GAT/ATC
Sac II	SstII, KspI	CCGC/GG
XhoI	PaeR7I	C/TCGAG

2.1.7 *Star Activity*

Under nonstandard conditions such as high pH, low ionic strength, high levels of organic solvents (for instance, glycerol or DMSO), a nonstandard ion in the reaction buffer (such as Mn^{2+} instead of Mg^{2+}), or even an elongated incubation period, enzymes may exhibit nonspecific recognition and cleavage, a phenomenon known as the **star activity**. In these cases, the enzymes will most likely cleave sequences, which differ by one or two bases from the canonical recognition motifs (Figure 2.5).

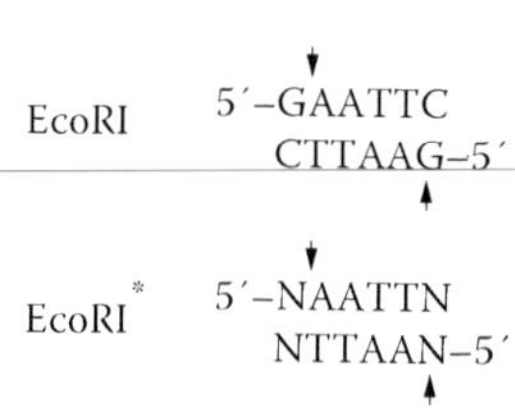

Figure 2.5 A schematic of star activity for the EcoRI example. The recognition specificity is compromised during star activity and the enzyme will cleave sequences that are similar but not identical to its canonical motif.

2.1.8 Restriction Mapping

The recognition motifs are quite specific under standard conditions, thus a DNA of a given sequence will be cut by a given restriction enzyme to give a unique set of fragments, while a different enzyme will generate a different set of fragments from the same DNA sequence. It is also possible to do multiple digest, either sequentially (first cut with the same enzyme, purify the resultant fragments, and then carry out the second digestion) or simultaneously (two different restriction enzymes can be used to digest the DNA molecule at the same time, if their buffer requirements are similar). In either case, the two restriction enzymes will digest from their respective recognition sequences, yielding a set of smaller fragments than single digests (Figure 2.6). These single and multiple digests can be applied to any DNA sequence to give a so-called restriction map of the sequence, even when the complete DNA sequence is not known.

2.1.9 Restriction Fragment Length Polymorphism

Our genome is largely variable within a population, resulting in polymorphisms. These polymorphisms can be detected by a variety of methods, however, one of the earlier methods to reliably detect a subset of these polymorphisms was restriction fragment length polymorphism, or RFLP. As the name implies, this method relies on whether or not polymorphisms among individuals cause a change in the restriction enzyme recognition motif or the fragment that digestion reaction produces, or not (thus it would not apply to polymorphisms, which do not create a readily detectable change in such recognition motifs or in the length of the restriction fragments).

Essentially, there are two major ways in which polymorphisms can create such changes: (a) a polymorphism (most likely a nucleotide substitution mutation) may demolish the recognition motif (Figure 2.7); (b) a polymorphism may be

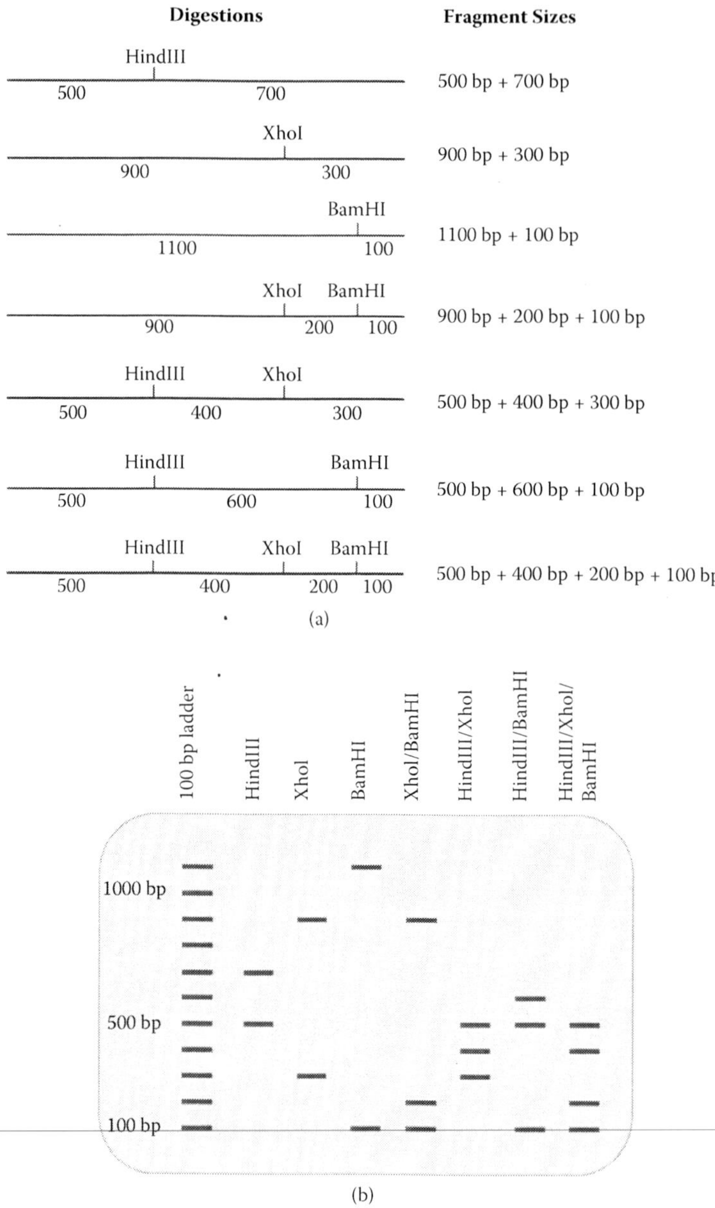

Figure 2.6 Restriction mapping principles. (a) A set of digestions conducted on the same DNA fragment. (b) This is how the digestion products would appear on the DNA agarose gel.

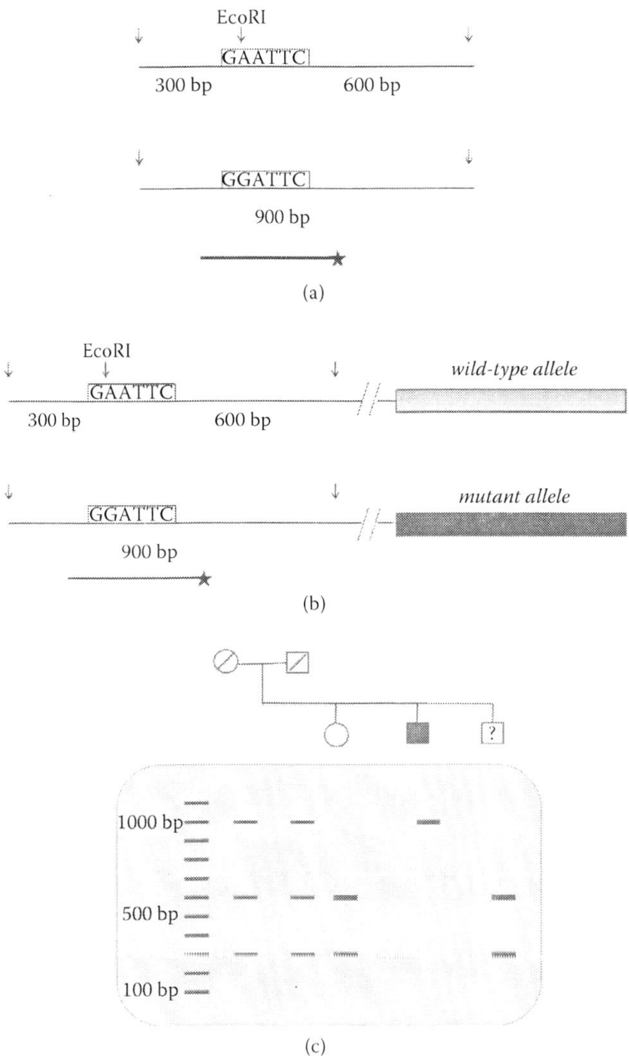

Figure 2.7 Restriction fragment length polymorphism (RFLP) principles. (a) A single nucleotide substitution changes the DNA sequence such that the EcoRI recognition motif is abolished, which can be detected by a labeled probe after restriction digestion. (b) If this polymorphism was closely linked to a gene related to a disease in the population, then the polymorphism can be used to indirectly detect the presence of a wild-type or mutant allele of the gene. (c) The RFLP can be used to trace the mutant allele, thus the genetic disorder, within a family (the question mark denotes the unborn child).

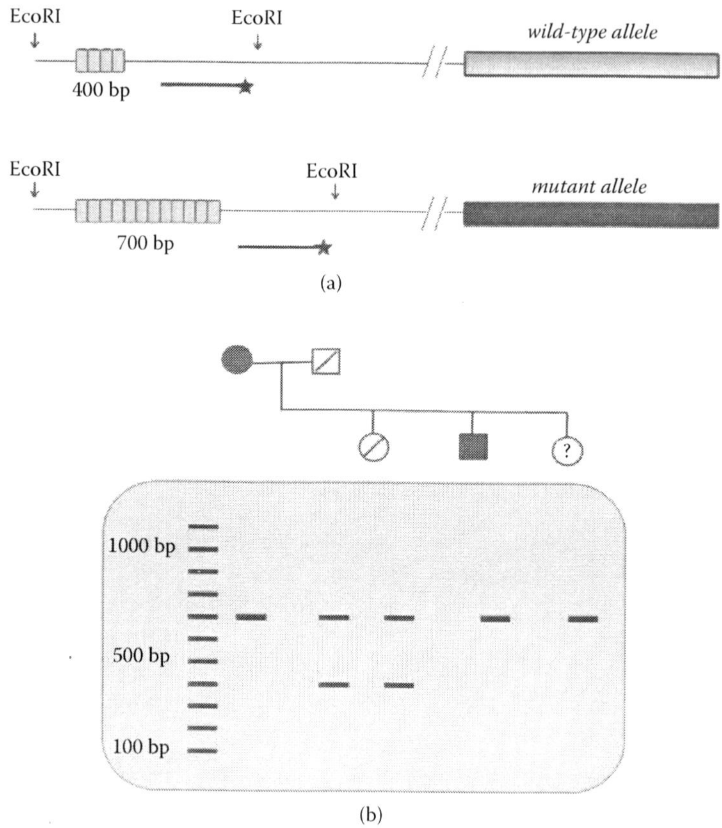

Figure 2.8 Restriction fragment length polymorphism (RFLP) due to a variable number of tandem repeats. (a) Different disease alleles could be linked to polymorphisms that result in different fragment lengths due to insertion of tandem repeats. (b) The fragment lengths can be analyzed by Southern blotting, using a labeled probe that can hybridize to both restriction fragments (the question mark denotes the unborn child).

due to an insertion or deletion of a stretch of DNA (a tandem repeat, a microsatellite, or similar) between two recognition sites (Figure 2.8). Under these conditions, the region of interest can be amplified and digested by the appropriate restriction enzyme, and analyzed by Southern blotting of the digestion products. If polymorphisms result in a change in the length of the digestion products, or restriction fragments, it would be possible to detect these changes using labeled probes.

RFLP maps of entire genomes can be prepared, but for any RFLP to be used, the locus has to be informative, which means the locus in question must be highly polymorphic across individuals in a population. Such RFLP maps of genomes could be used to study genetic diseases to a certain extent, but because of the recombination frequencies between various RFLP loci, new and more improved techniques have been developed to isolate novel genes.

2.2 Vectors

Vectors are DNA molecules which can replicate autonomously and thus can be used to carry the insert DNA into organisms and amplify this DNA *in vivo*. There are many types and many functions of vectors. The most commonly used vectors that will be covered in this section are (a) plasmids, (b) phage vectors, (c) cosmids, (d) bacterial artificial chromosomes, and (e) yeast artificial chromosomes. All of those vectors change in the size of the insert they can carry, and the purpose for which they can be used. Plasmids, for example, can carry inserts of up to 10 kb, while phage vectors go up to 20 kb inserts, and YAC vectors can carry 100 to 1000 kb inserts (Table 2.2). One can also choose vectors based not on the size of the insert, but on the application purpose: cloning, sequencing, preparing RNA or DNA probes, or expressing proteins (Hartl et al. 1988; Howe 2007; Nair 2008).

2.2.1 Plasmids

Plasmids are perhaps the most commonly used vectors and are extrachromosomal DNA molecules that are present in prokaryotes, and offer a wide range of functions from production of conjugation pili (F plasmids), conferring antibiotic resistance

Table 2.2 Various Commonly Used Laboratory Vectors and Their Key Features

Vector Type	*Insert Size*	*Examples*	*Purposes*
Plasmid	10–20 kb	pUC19, pCMV	DNA manipulation; protein expression; and many others
Phage (λ, insertion)	Around 10 kb	λ gt11	cDNA libraries
Phage (λ, replacement)	Around 23 kb	EMBL4	Genomic DNA libraries
Cosmid	Around 45 kb	pHM1; pJB8	Genomic DNA libraries
Phagemid	10–20 kb	pBluescript	DNA manipulation; *in vitro* transcription; *in vitro* mutagenesis
BAC	130–150 kb	pBACe3.6	Genomic DNA libraries
YAC	1000–2000 kb	pYAC4	Genomic DNA libraries

(R plasmids), sugar fermentation, heavy metal resistance, and so on, depending on the genes expressed on the plasmid. Plasmids are usually small, circular double-stranded DNA molecules that have the capacity to replicate autonomously within bacteria; however, replication is still coupled to host replication and can be found in two forms: *stringent*, which replicate once or twice per generation (**low copy number plasmids**), and *relaxed*, which replicate 10–200 copies in each generation (**high copy number plasmids**) (Table 2.3). For use in cloning purposes, all plasmids must contain the following DNA sequences:

Table 2.3 Copy Numbers of Some Key Plasmids

Plasmid	*Plasmid Size (Approx.) (bp)*	*Copy Number*
pUC	2700	500–700
pBR322	2700	>25
ColE1	4500	>15

1. **Origin of replication** (also called ***ori***), is required for autonomous replication within the host cell. If the host is a bacterium, then a bacterial ori is required; if the host is yeast, then a yeast ori is used.
2. **Selective marker** is required for selection of recombinant bacteria that contain the plasmid; the most common markers are antibiotic resistance genes.
3. **Multiple cloning site** (**MCS**) is an engineered sequence of DNA that contains multiple restriction sites that are unique (i.e., that are not found anywhere else in the plasmid), which will be used for inserting the foreign DNA.

In addition to these three basic properties expected of plasmids, one can also expect to find a host-compatible promoter for expression vectors, and RNA termination sequences, or an MCS inserted within the *beta*-galactosidase (lacZ) gene to be used for blue-white screening (as discussed in Section 2.4.2).

Most of the commonly used laboratory plasmids today are based on the naturally occurring *E. coli* plasmid ColE1; but these naturally occurring plasmids have the disadvantage of not being too flexible in the unique sites they have which can be used for cloning, and the difficulty of selecting recombinants. The first vector that gained widespread laboratory use was the pBR322 plasmid, which was developed by Paco Bolivar and Ray Rodrigues (Bolivar et al. 1977). While screening for bacteria with ColE1, plasmids relied on the resistance gene on the plasmid, which helped bacteria escape the lethal

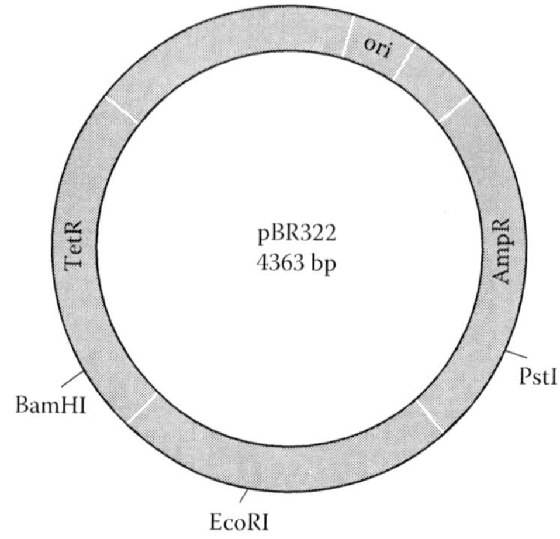

Figure 2.9 The simplified plasmid map of the pBR322 vector. The *ori* sequence is derived from the ColE1 plasmid, and, in addition, the plasmid contains two antibiotic resistance genes that can be cleaved by two different restriction enzymes and used for cloning.

effects of the drug colicin. More recently, simple antibiotic screening methods are used to screen cloning experiments with pBR322, as well as other plasmids. The pBR322 plasmid contains two genes that can be used for selection: the ampicillin resistance gene and tetracycline resistance gene (Figure 2.9).

These antibiotic selection genes, namely *beta*-lactamase (*bla*) for ampicillin resistance, and the tetracycline/H+ antiporter gene (*tet*) for tetracycline resistance, each have one unique enzyme recognition motif within their coding sequences, such that when an insert DNA is cloned into the PstI site of the AmpR gene, for example, then the bacteria harboring this recombinant plasmid will lose its ability to grow on ampicillin selection but retain its tetracycline resistance. Through replica plating, the bacteria on an ampicillin-containing agar plate and another set on a tetracycline-containing agar plate, one can monitor whether the bacteria transformed with the plasmid have the insert DNA or not (Figure 2.10).

The next generation of plasmids was engineered to incorporate different methods for selection. The pUC series of

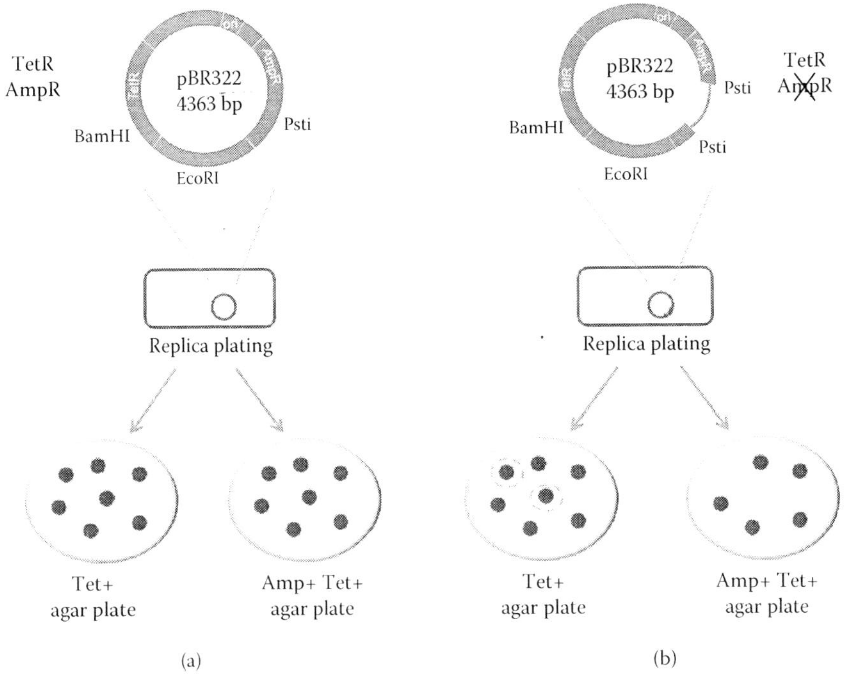

Figure 2.10 A schematic summary of a cloning strategy using the pBR322 vector. Essentially, transformed bacteria are replica plated onto two different agar plates—one containing ampicillin as a selection marker, and the other containing tetracycline. (a) If the plasmid vector contains no insert, then both resistance genes are intact, and bacteria will grow on both agar plates. (b) If the plasmid vector contains an insert DNA cloned into, for instance, the ampicillin resistance gene, then the bacteria will grow colonies on the Tet+ agar plate, but not on the Amp+ plate.

plasmids (so named because they were developed at the University of California) contain not only an antibiotic selection, but also another selection strategy, which combines an engineered MCS within the coding sequence of the *beta*-galactosidase gene (Figure 2.11). (The selection strategy for this vector will be described in detail in Section 2.4.2.)

One of the most significant properties of this plasmid is the engineered MCS sequence, which contains multiple unique restriction enzyme recognition motifs. As with the pBR322

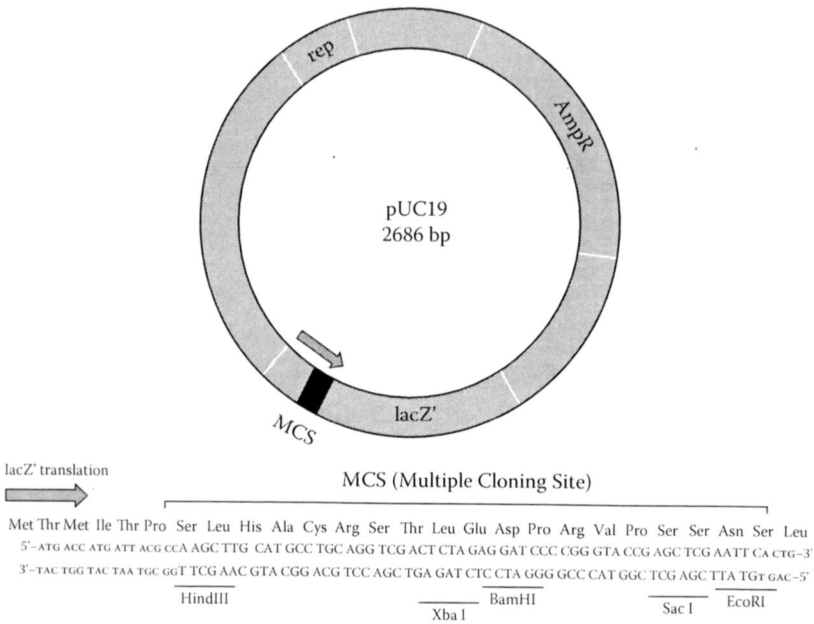

Figure 2.11 The simplified plasmid map of the pUC19 vector. The *rep* sequence is derived from the pMB1 replicon and confers an autonomous replication property to the plasmid; the *AmpR* gene allows for ampicillin selection; additionally the plasmid is engineered to harbor an MCS within the lacZ′ coding sequence (only some of the recognition sequences within the MCS are shown).

plasmid, the ampicillin resistance gene can be used for selection, however, a second selection can be carried out due to the insertion of the MCS within the coding sequence of the lacZ' gene (codes for the first 63 amino acids of the lacZ α-peptide). The presence of an insert DNA within the MCS will disrupt the lacZ α-peptide function, but will not affect the ampicillin resistance gene.

There are a variety of different plasmids that have been generated since that time, and a wide range of markers that can be used depending on the organism that the plasmids will be transferred into (*E. coli*, yeast, *Drosophila*, plants, mammalian cells, etc.). Some of the specialist purpose plasmids will

Table 2.4 Some Common Selectable Markers Used for Cloning and Their Modes of Action

Marker	*Acts On*	*Mode of Action*
Ampicillin	Prokaryotes (gram negative bacteria)	Inhibits cell wall synthesis, thus bacteria cannot replicate in the presence of ampicillin
Tetracycline	Prokaryotes	Binds to the 30S ribosomal subunit and inhibits translocation of ribosomes
Kanamycin	Prokaryotes and eukaryotes	Binds to ribosomal subunits and inhibits protein synthesis
Puromycin	Prokaryotes and eukaryotes	Binds to the ribosomal A site and causes premature chain termination
Cycloheximide	Eukaryotes	Targets the E site of the 50S subunit of eukaryotic ribosomes

be covered in the following sections, and the most commonly used selectable markers are listed in Table 2.4.

Some antibiotics, such as ampicillin, target bacterial cell wall synthesis, and thus are only active against bacteria, while others, such as puromycin is an aminonucleoside that mimics the 3′ end of the aminoacyl-tRNA molecule, and thus enters the A site of the ribosome and causes premature termination of protein synthesis: it is therefore effective on both prokaryotic and eukaryotic ribosomes (Table 2.4).

Stable maintenance of plasmids in bacteria relies on a partitioning system, the *par* region, which ensures almost equal segregation of the duplicated plasmids to each daughter bacteria after fission. This is particularly important for low copy number plasmids, otherwise plasmids may become lost from the bacteria. Therefore, such *par* regions can be cloned into plasmids of low copy number to ensure stability across generations.

If two plasmids contain the same *par* regions, or use the same replication mechanism, then the two plasmids are said to be **incompatible**, that is, they cannot coexist in the same bacterium in the absence of selection pressure. Thus, if bacteria are to be transformed with two different plasmids simultaneously for the purposes of the assay, it would be safe to use two different selection markers.

2.2.2 Phage Vectors

Plasmids used to be fairly common in a typical molecular biology laboratory, particularly if one is to study larger DNA fragments, such as for cDNA or genomic DNA libraries. However, with the advancement of the technologies available, phage-based vectors are no longer as commonplace as they used to be. Still, from a historical perspective, we will present a short overview of phage vectors.

Initial studies on the phage life cycle and genome (Figure 2.12), which date back to the 1950s, and in particular to the work of Lwoff and his coworkers and followed more recently by Ptashne and his group in the 1990s, have shown that phages are convenient alternatives to plasmids due to the 50 kb genome that becomes packaged into the head region, and the central part of this genome is not necessary for packaging and thus can be replaced with an insert DNA.

Wild-type λ phage itself cannot be directly used for cloning purposes; first, it contains very few unique restriction sites that can be exploited for cloning, and second, there is a maximum size limit (between 78% and 105% of wild-type DNA length, or 37–53 kb) to the DNA that can be packaged to the phage heads. Therefore, the phage genome (Figure 2.13) had to be engineered before it could be used as a cloning vector. One feature is that the genes required for recombination can be removed, and a lytic cycle can still take place; another issue is the removal of certain restriction enzyme recognition sequences without disrupting gene function (through genetic

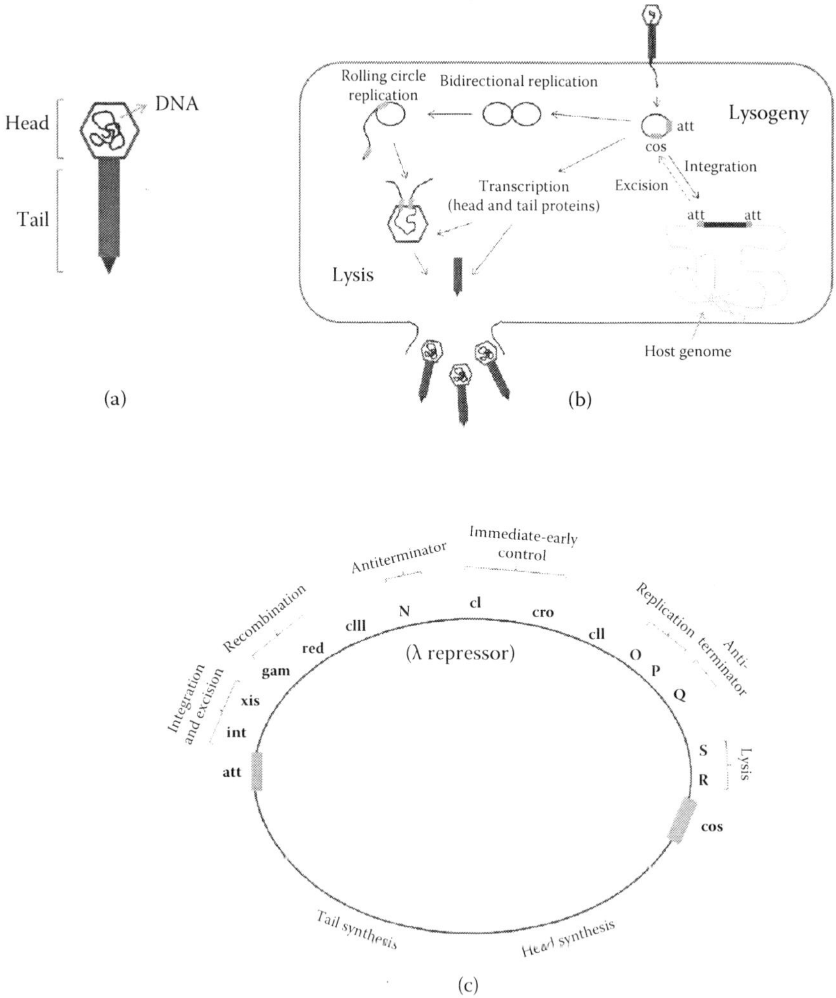

Figure 2.12 An overview of the l phage structure, life cycle, and genome. (a) The simplified structure of the l phage, consists of the phage genome and the head and tail proteins. (b) A schematic representation of the life cycle of the l phage. Upon infecting the host, the phage can assume either one of the two fates: it can either integrate its own DNA into the host genome and lysogenize, or it can replicate, express head and tail proteins, assemble into new phage particles, and lyse the cell. (c) Representation of the phage genome, where some of the genes that are important for the phage life cycle are shown. cI and cro repressors are major regulators of phage transcription.

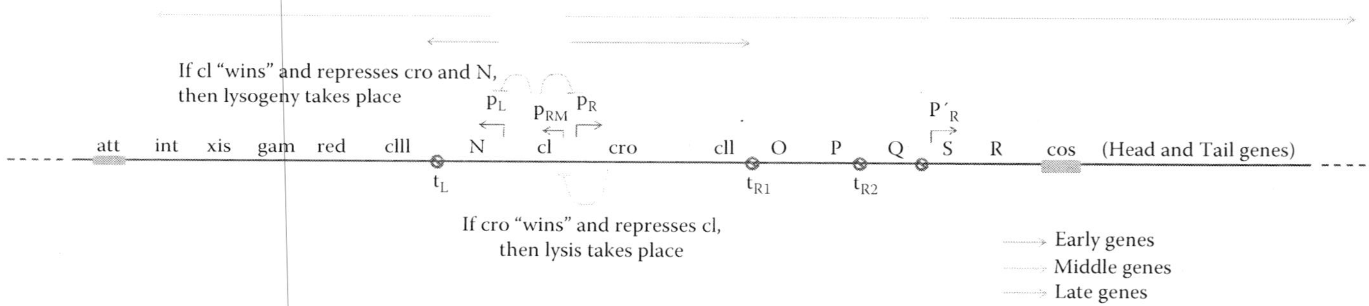

Figure 2.13 **Summary of the temporal control in l phage transcription. Early transcription occurs from two promoters, pL and pR (expression of the cl repressor itself is maintained through a third promoter, pRM) and stops at the termination signal sequences, tL and tR1 (a low level transcription can carry on until tR2). The product of gene N, namely pN, is an antiterminator which allows the RNA polymerase to read beyond tL and tR2, which is when the l phage switches to middle gene transcription. Genes to the left of N are the genes involved in recombination, and the genes to the right of tR2 are those required for phage DNA replication. At this stage, if sufficient levels of the cro product accumulate in the cell, it represses transcription of the cl gene and early transcripts, then at the same time the product of gene Q, which is another antiterminator, builds up and allows for the transcription from p′R, S and R genes required for lysis, followed by transcription of the remaining late genes that code for head and tail proteins. However, if the cl gene product accumulates in the cell, it represses the cro gene and does not allow for transcription of the late genes, and drives the phage toward lysogeny. In times of danger, such as DNA damage to the host cell, the cl gene product, or the l repressor, gets cleaved, leading to the accumulation of the cro repressor and ultimate lysis.**

engineering—which will be covered in other sections), thus new vectors were developed based on bacteriophages.

There are two types of phage vectors: (i) **insertional vectors** and (ii) **replacement vectors**. Insertional vectors are derived from wild-type λ phage and have one or more unique site(s) (MCS) to which insert DNA can be cloned; in some vectors, this MCS could reside within an engineered lacZ coding sequence (Figure 2.14a). Replacement vectors, on the

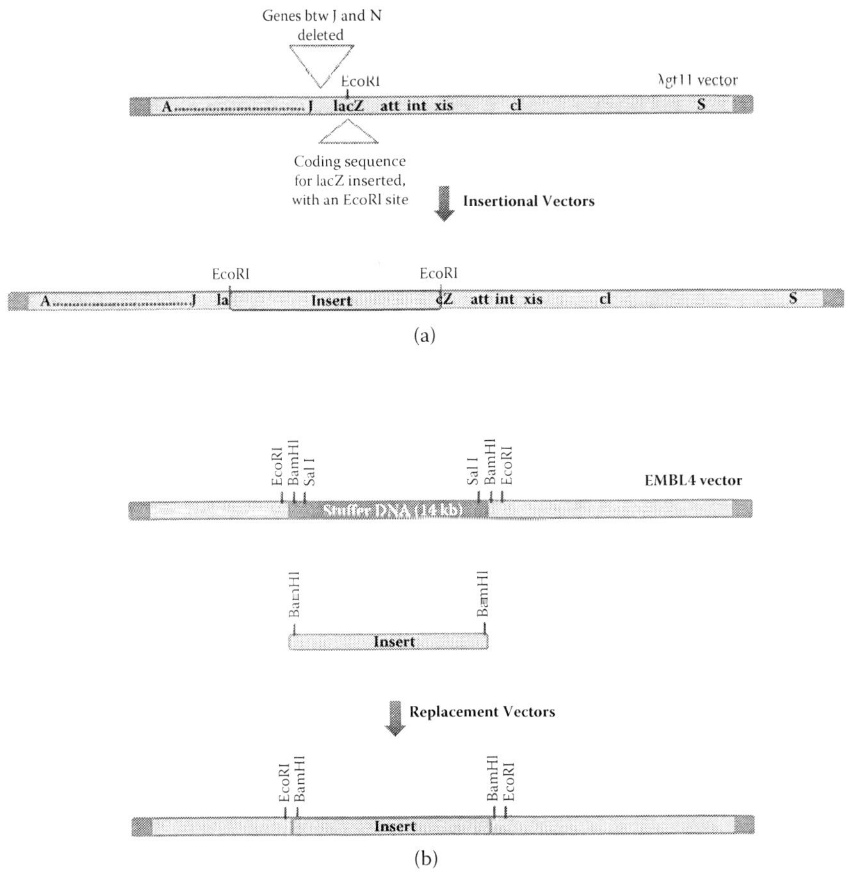

Figure 2.14 Examples of (a) an insertional vector, and (b) a replacement vector. (a) In insertional vectors, the insert DNA is cloned into the MCS, which in this case resides within the lacZ coding sequence. (b) In replacement vectors, there is a central stuffer for proper packaging of the vector, which will be "swapped" with the insert DNA.

other hand, do not contain an MCS into which the insert can be cloned, but rather have a stuffer region flanked by a few restriction enzyme recognition sequences that can be used to "swap" the insert with the stuffer DNA (Figure 2.14b).

After cloning, the phage particles will be prepared through *in vitro* packaging, and will be incubated with bacteria for screening of recombinants (Figure 2.15). The phage will lyse the bacteria upon incubation, and will yield clear plaques, which is commonly known as the **plaque assay**.

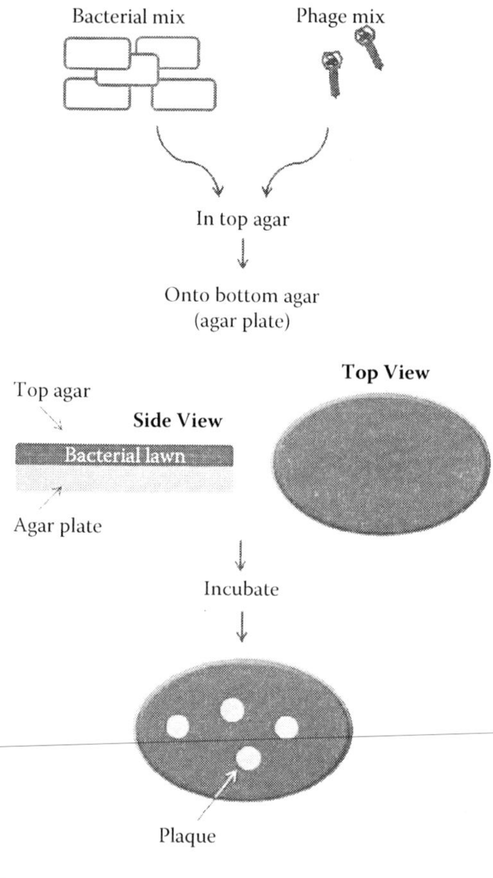

Figure 2.15 The plaque assay. Bacteria and phages are mixed together in the top agar, and then poured on top of an agar plate (the bottom agar), where the bacteria grows into a "bacterial lawn," and after incubation and growth of the phage, the lysed bacteria will yield a clear plaque.

However, regardless of whether one uses insertional or replacement vectors for cloning purposes, packaging the recombinant phage vector into phage particles is a difficult task, mainly because a large portion of the genes required for head and tail assembly are deleted in either vector. To overcome this problem, one simply provides these proteins from lysates of bacteria lysogenic with the defective phage (Figure 2.16). (Otherwise maintaining those bacteria in stock would have been a problem in its own right!)

2.2.3 Cosmids and Phagemids

Cosmids are essentially plasmid vectors that contain λ phage *cos* sites, developed by Collins and Bruning (1978). Since they are based on plasmids and have an origin of replication, they can replicate in the cell like a plasmid, but because they have *cos* sites they can be packaged like a phage, leading a *dual life*. And since they can be packaged like a phage particle, they may even at times carry up to 45 kb inserts, a much larger capacity than a common plasmid, and even larger than a typical λ phage vector that can carry only up to 25 kb. Just like the phage vectors, cosmids are also not as commonplace as they used to be; more advanced *specialist* vectors are now in use (Figure 2.17).

Phagemids, similarly, are not as common as they used to be, but in many laboratories one can still come across the classical phagemid, pBluescript. Phagemids are essentially plasmids that contain an origin of replication for single-stranded phages (such as f1), thus bacteria which are transformed with this plasmid and infected with a helper phage (such as M13 or f1) can produce single-stranded copies of plasmids, which in turn can be packaged into phage heads. In the absence of a helper phage, the DNA would be propagated like a normal plasmid in bacteria. Since this vector is a hybrid of both plasmids and single-stranded phage vectors, it can be used to generate

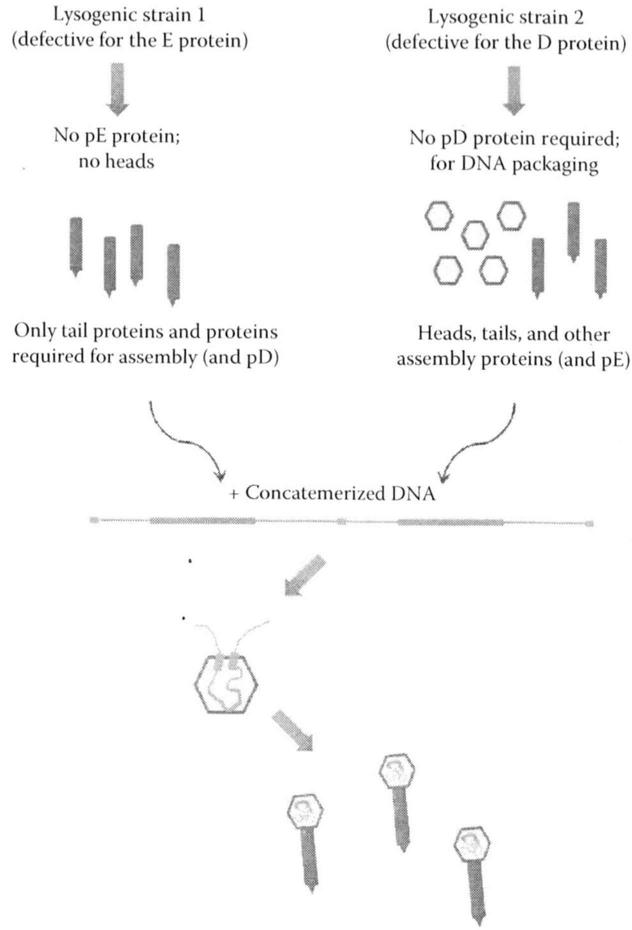

Figure 2.16 ***In vitro*** **packaging of recombinant λ phages. Two bacterial strains lysogenic with phages defective in either head protein production (strain 1) or DNA packaging (strain 2) are used to synthesize the head, tail, and assembly proteins required for packaging of DNA. Mixing the lysates of these two strains with the concatemerized recombinant DNA will result in the recombinant phage vector DNA being packaged properly into phage particles. These phage particles can then be screened for true recombinants.**

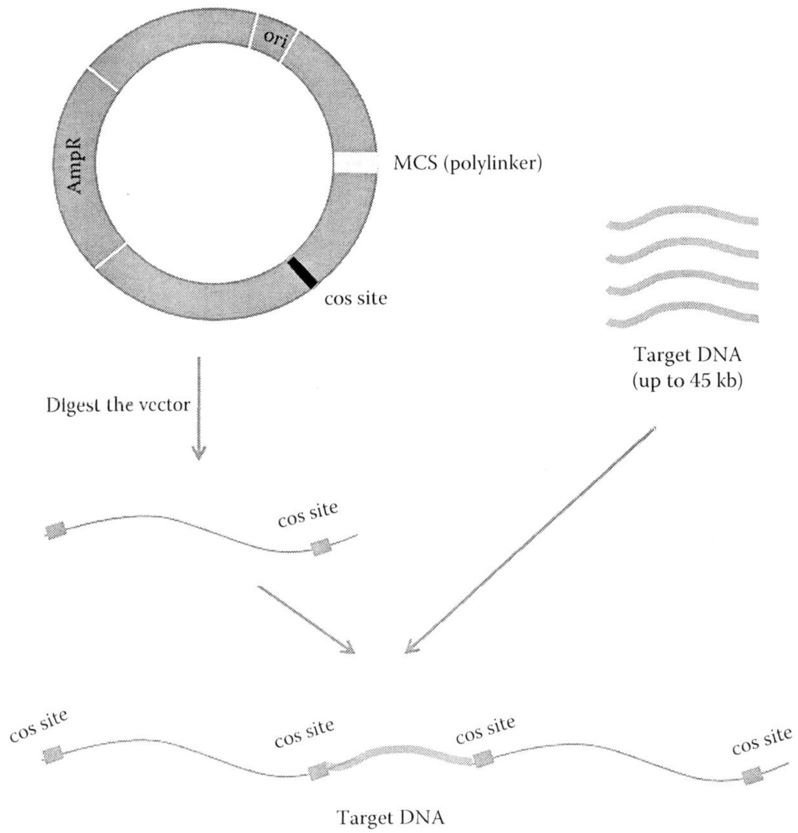

Figure 2.17 A schematic of a typical cosmid vector, showing the major elements: an origin of replication (ori), a selectable marker (such as the ampicillin resistance gene), a polylinker or MCS, and a cos site. Upon digestion of the vector within the polylinker, target DNA is cloned into the linearized vector, which can then be packaged into phage particles.

single-stranded DNA to be used in sequencing reactions. Today, the most typical phagemid that can be found in molecular cloning laboratories is the pBluescript phagemid variants, used for cloning, sequencing, site-directed mutagenesis, or *in vitro* transcription purposes (Figure 2.18).

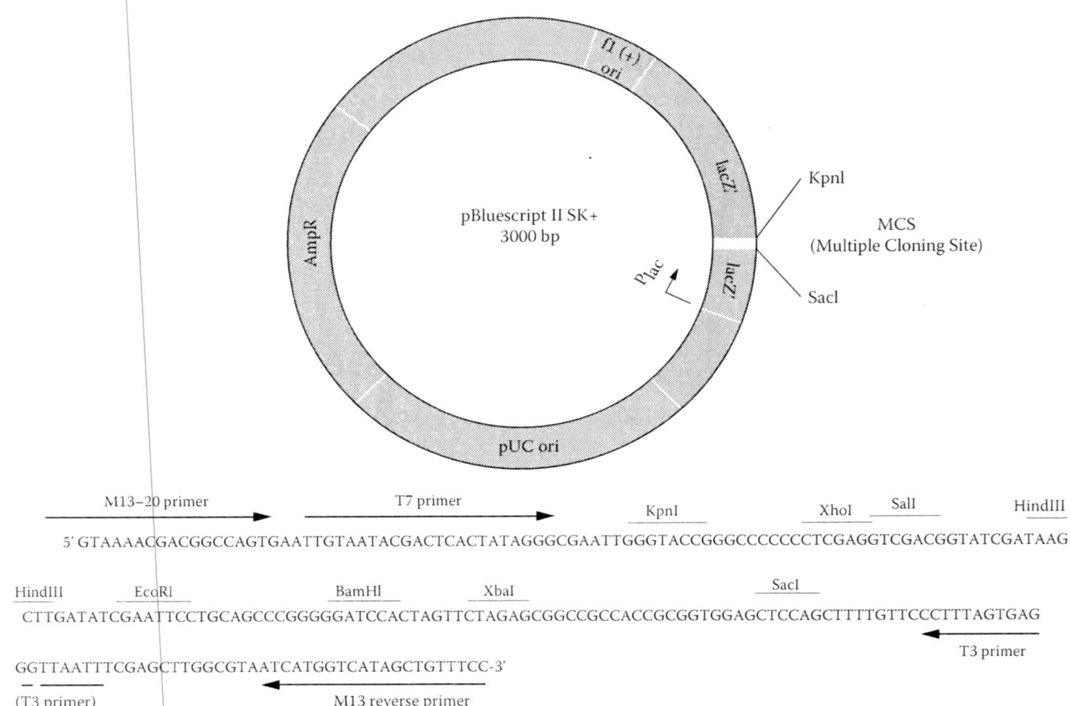

Figure 2.18 A schematic diagram of the pBluescript II SK+ phagemid, with the ampicillin resistance gene, multiple cloning site (MCS), f1 origin (in + orientation, hence SK+), and the pUC ori sequence. The base sequence of MCS is given below the map, with the main restriction sites indicated above the sequence.

2.2.4 Specialist Vectors

As recombinant DNA technology has advanced, researchers have generated modified vectors for their own specialized purposes. In this section we will only concentrate on a few of these, as the number and variety of specialist vectors change with the model organism and field of interest. We will mainly discuss bacterial artificial chromosomes (BACs), yeast artificial chromosomes (YACs), expression vectors including pGEX series or pCMV-based plasmids, and vectors coding for enhanced green fluorescent protein (pEGFP) as examples. It should be emphasized again that this is a very narrow selection and does not by any means imply that these are the most common expression vectors used—on the contrary, quite a significant number of *Drosophila* vectors, plant vectors, and vectors for expression in other organisms have been omitted in order to keep the chapter as concise as possible. These vectors merely serve the purpose of practical examples.

2.2.4.1 Bacterial Artificial Chromosomes

Bacterial artifical chomosomes (BACs) are based on the Fertility (F) plasmid that is designed to carry large DNA sequences of usually 150 to 350 kb, mainly for genome library construction purposes. They are based on F plasmids, because the *par* genes ascertain the even distribution of large-sized recombinant plasmids to the next generation.

A typical BAC vector contains ori and rep sequences to ensure replication of the vector and the copy number, and par sequences for even partitioning of the DNA to daughter cells, in addition to a selectable marker and usually phage promoters such as the T7 promoter for transcription of the cloned genes (Figure 2.19).

BAC vectors have been the vectors of choice for genomic sequencing projects due to their stability (the probability or rearrangement of the large-sized insert is minimal), hence they are often used to construct BAC libraries. These genomic BAC

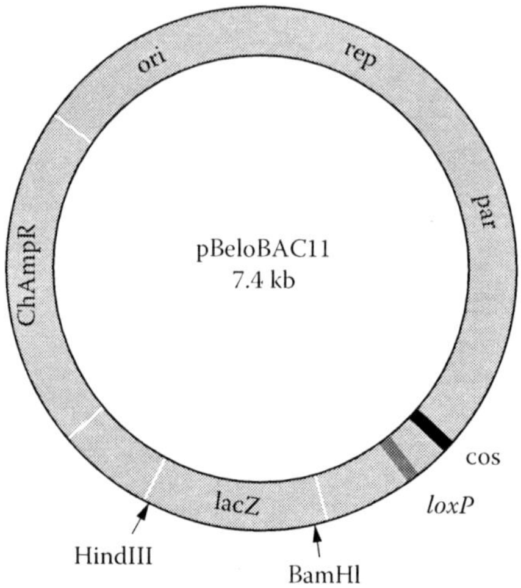

Figure 2.19 A schematic diagram of the pBeloBAC11 vector. The vector contains the chloramphenicol resistance gene as a selectable marker, and ori sequence, rep and par sequences, loxP site and cos site, in addition to a lacZ gene that includes two restriction recognition motifs for cloning (HindIII and BamHI).

libraries can be used in combination with transgenic mice, particularly since BAC transgenes were found to express at appropriate developmental timing as endogenous genes in transgenic models and reduce positional effects (Beil et al. 2012; Van Keuren et al. 2009) (see Chapter 8, Section 8.2 for transgenic animals).

2.2.4.2 *Yeast Artificial Chromosomes*

As the research of the human genome began, even before there was an official Human Genome Project, researchers had started obtaining physical data about the organization of chromosomes, however, lambda or cosmid vectors can usually contain up to 5 or 10 kb of DNA, which posed a serious problem. **Yeast Artificial Chromosomes (YACs)** have come to the rescue, as they could be used to accommodate up to 100 kb of DNA fragments, and thus increase the likelihood of containing

Table 2.5 Similarities and Differences between BAC and YAC Vectors

	BAC Vector	*YAC Vector*
Mode	Circular	Linear
Copy number	1–2 per cell	1 per cell
Capacity	100–350 kb	Almost limitless
Stability of insert	Stable	Unstable

a DNA region that exists only as a single copy among the chromosomes, thus reducing the possibility of misalignments of clones. Recently, the maximum size of an insert that can be accommodated by a YAC is almost limitless (see Table 2.5).

The other important property of YACs is that they mimic normal chromosome structure (hence the name *artificial chromosomes*, within BACs), as they consist of two *arms* between which large DNA fragments can be cloned. Each arm contains a telomere at the end for stabilization like a normal chromosome, and one of the arms contains an autonomous replication sequence (ARS) required for yeast chromosome replication, a centromere (CEN), and a marker for selection of recombinant yeast (usually a gene for the synthesis of an amino acid, such as *trp1* or *his3*) (Figure 2.20).

The cloned vector is used to transform yeast that is mutant for *trp1* and *ura3*, and if red colonies appear that implies that yeast cells contain both telomeres due to complementation, and the cells appear red due to inactivation of a vector DNA if there is an insert.

2.2.4.3 *Expression Vectors*

If one needs to synthesize specific proteins for further analysis, there are several expression vectors that one can choose from, depending on which cell type from which organism will be used for gene expression. Unfortunately, we cannot cover

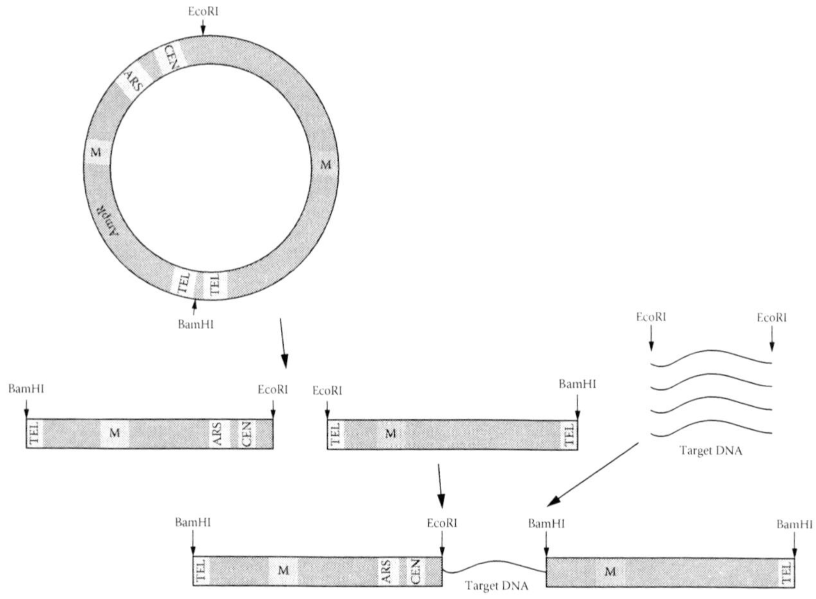

Figure 2.20 A diagram of a generic YAC vector. The vector contains two yeast markers (M), ampicillin resistance gene (AmpR), telomere sequences (TEL) separated by a BamHI site, centromere sequence (CEN), and autonomously replicating sequence (ARS) for replication in yeast. The vector is digested by both the EcoRI and BamHI enzymes, which generates two linear fragments; in the meantime, the insert DNA is prepared by EcoRI digestion of target DNA. All the fragments are then allowed to ligate, and the recombinant artificial chromosome is generated, with the insert carried in between the short and long arms from the YAC vector.

every expression vector. that is available, nor can we cover every host organism. This section merely intends to provide readers with a basic idea of how to work with such plasmids; therefore, only three different examples have been selected.

pBluescript vectors (see phagemids in Section 2.2.3, "Cosmids and Phagemids") can be used for *in vitro* transcription and translation, due to the presence of T7 and T3 phage promoters on either side of the polylinker (Figure 2.18) (also see Chapter 4, "Protein Production and Purification"). However, if one would like to express a protein in a bacterial cell, a plant cell, or a mammalian cell, one needs to use a plasmid that contains either a bacterial, plant, or mammalian promoter, respectively, upstream of the gene of interest.

pGEX series of vectors are commonly used for bacterial expression of proteins in large quantities. These vectors express the gene of interest as a fusion to the glutathione-S-transferase (GST) gene driven by the *tac* promoter, a hybrid of *trp* and *lac* promoters (Figure 2.21). The promoter is upstream of the GST gene, followed by a polylinker (or MCS) where the gene of interest will be cloned **in frame** with the GST, and this GST fusion will be used for a variety of purposes (see Chapter 4, "Protein Production and Purification" and Chapter 6, "Protein–Protein Interactions").

For expressions in mammalian cells we will discuss one of the most commonly used plasmids, the pCMV series, although many other vector types exist. These vectors typically rely on the expression of the gene of interest from a promoter that strongly activates transcription in mammalian cells—since viruses are very good hijackers of the transcriptional apparatus of their host cells, strong promoters from viruses that infect mammalian cells, such as the cytomegalovirus (CMV) or simian virus 40 (SV40) are usually found in these vectors. In the case of pCMV, it is the CMV promoter (Figure 2.21). In the particular example given (pCMV-HA), the promoter is upstream of a hemagglutunin (HA) tag followed by an MCS (Figure 2.21). The HA tag can be used for a variety of purposes, which will be discussed in the following chapters (see Chapter 4, "Protein Production and Purification").

Many of the more recent and commonly used plasmids will also be discussed in other chapters.

2.3 Modifying Enzymes

Apart from the restriction-modification system that has already been covered, many different enzymes that modify DNA

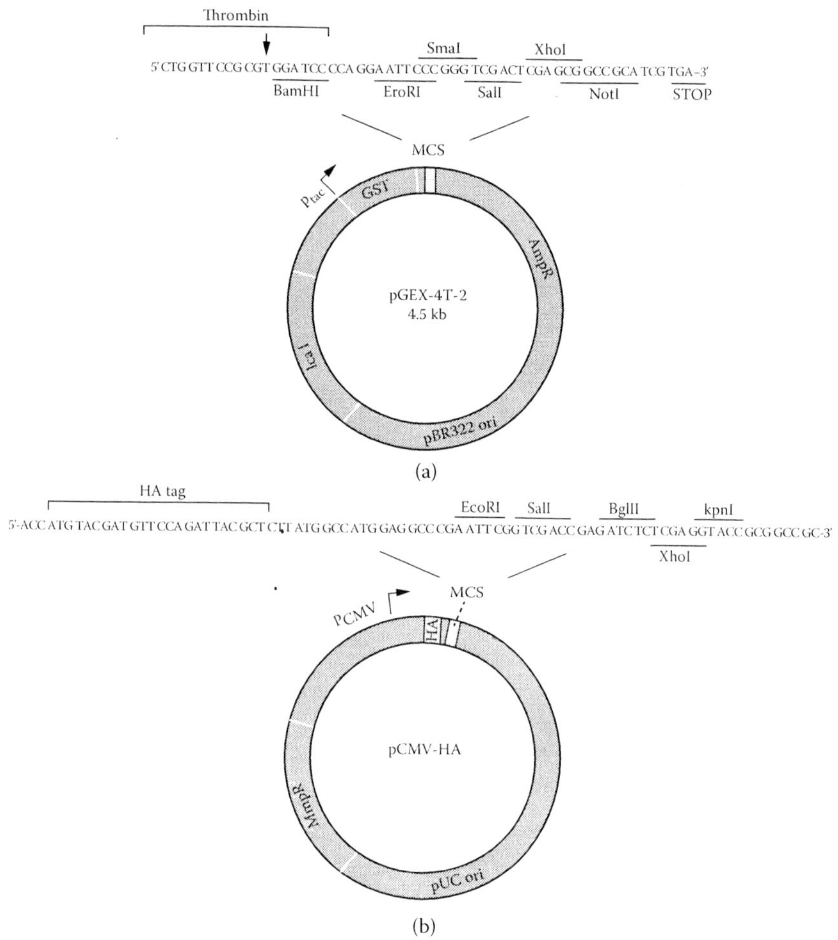

Figure 2.21 Diagrams of pGEX-4T-2 and pCMV-HA vectors. (a) pGEX vectors typically contain an antibiotic resistance gene, an ori sequence, a bacterial promoter, in this case the hybrid *tac* promoter, driving the expression of a glutathione-S-transferase (GST) gene followed by a polylinker or MCS sequence, where the gene of interest will be cloned for bacterial expression. (b) pCMV vectors typically contain an ori sequence, antibiotic selection for cloning, and a cytomegalovirus (CMV) promoter upstream of an HA tag followed by an MCS.

molecules exist. In this section only some of these that are routinely employed for cloning applications will be discussed: **polymerases**, which are frequently used to amplify the target DNA to be cloned; **ligases**, which are used for joining two different DNA fragments and thus for cloning; and **alkaline phosphatases** that are used to phosphatase the 5′-phosphate groups on vector DNA to be used in cloning. Other modifying enzymes such as nucleases, reverse transcriptases, and so on, will be discussed in other chapters.

2.3.1 Polymerases

DNA polymerases are DNA-dependent DNA polymerases, that is, they use a DNA strand as a template and synthesize a complementary DNA molecule in a 5′-to-3′ direction, also called the *5′-3′-polymerase activity*. DNA polymerases, however, cannot start *de novo* synthesis, that is, all DNA polymerases require an existing free 3′-OH group, routinely provided by **primer** sequences in molecular biology applications, to start a synthesis reaction, due to their proofreading activities (also called *3′-to-5′-exonuclease activity*) (Figure 2.22).

Historically, the most frequently used DNA polymerase for genetic engineering purposes was the *E. coli* DNA polymerase I, which is composed of three major domains with different catalytic activities: 5′-3′-polymerase activity, 3′-5′-exonuclease activity, and 5′-3′-exonuclease activity. The 5′-3′-exonuclease activity of the *E. coli* enzyme could interfere with certain experiments and is usually not desired for molecular biology applications. The **Klenow fragment**, which is commonly used in a variety of molecular biology applications, is produced by proteolytic cleavage of the *E. coli* DNA polymerase, and possesses only 5′-3′-polymerase and 3′-5′-exonuclease activities, and thus it can be used more readily in research.

One other modifying enzyme used in research that is similar to the Klenow fragment is the T4 DNA polymerase of bacteriophage T4, and similar to the Klenow fragment, it lacks

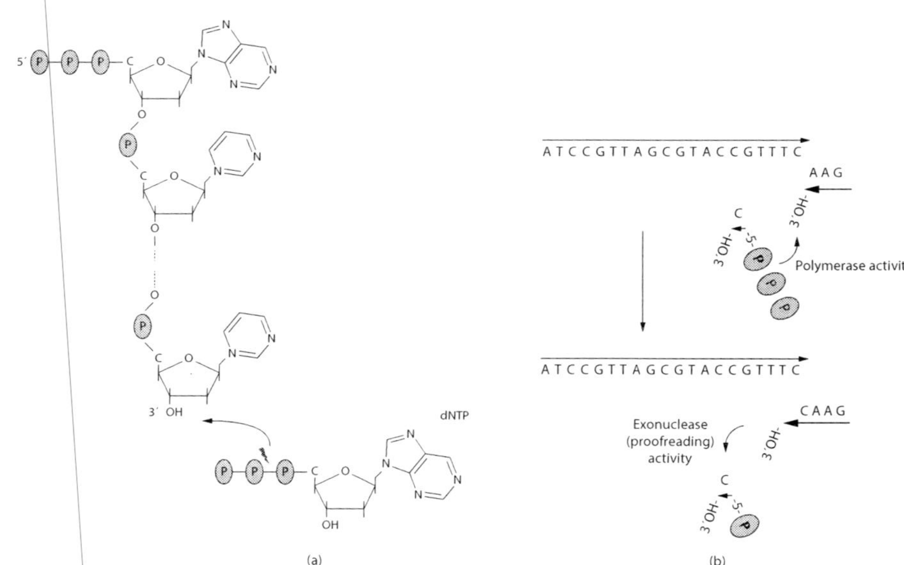

Figure 2.22 A simplified diagram summarizing the mode of action and the catalytic activities of DNA polymerases. (a) DNA polymerase 5′-3′-polymerase activity is responsible for covalently linking the 5′-phosphate group of an incoming free deoxynucleotide (dNTP) to the 3′-OH group of an existing primer sequence, using the energy released from breaking the high-energy phosphate bonds in the dNTP monomer. (b) The DNA polymerase uses a DNA template strand to generate the complementary molecule, but requires the presence of a primer sequence to initiate synthesis—this primer is essential for the 3′-5′-exonuclease activity (or proofreading activity) of the enzyme, whereby an incorrectly inserted mismatched nucleotide will be removed.

the 5′-3′-exonuclease activity. Thus, T4 DNA polymerase can be used in almost all of the reactions that the Klenow fragment is used; however, its exonuclease activity is much higher than the Klenow fragment and therefore it is the preferred enzyme for digesting 3′ overhangs so as to generate blunt-ended DNA molecules.

For applications such as a polymerase chain reaction (PCR), where DNA double strands are melted at a high temperature (see Appendix A, Figure A.3), the above-mentioned DNA polymerases will not be too practical since they are inactivated at high temperatures. The report of a heat-stable DNA polymerase from *Thermus aquaticus*, a bacterium that lives in hot springs, in 1976 came as a relief to the biotechnology community almost a decade later for PCR applications (Chien, Edgar, and Trela 1976). The reaction catalyzed by this enzyme is essentially the same, apart from the fact that magnesium is not required as a cofactor, and the fact that this enzyme retains catalytic activity at 75–80°C, its half-life at 95°C is approximately 1.5 hours, and it is mostly not active at 37°C. Several other variants have been reported since then, and are commercially available for researchers: *Pfu* polymerase, for example, has been isolated from *Pyrococcus furiosus* and has a much lower rate of replication errors.

2.3.2 Ligases

DNA ligases catalyze the covalent bond formation between a 5′-phosphate group and a 3′-hydroxyl group, however, unlike DNA polymerases, they cannot join a free deoxynucleotide to an existing primer sequence; instead, they act as DNA repair enzymes and “seal” the “nicks” of double-strand breaks in nucleic acids (Figure 2.23). The DNA strands to be joined by DNA ligases usually result from hydrolysis of an existing sugar-phosphate backbone, therefore the 5′-end of one of the strands has only one phosphate group, instead of three phosphates in dNTPs. Therefore, ligases require an external energy

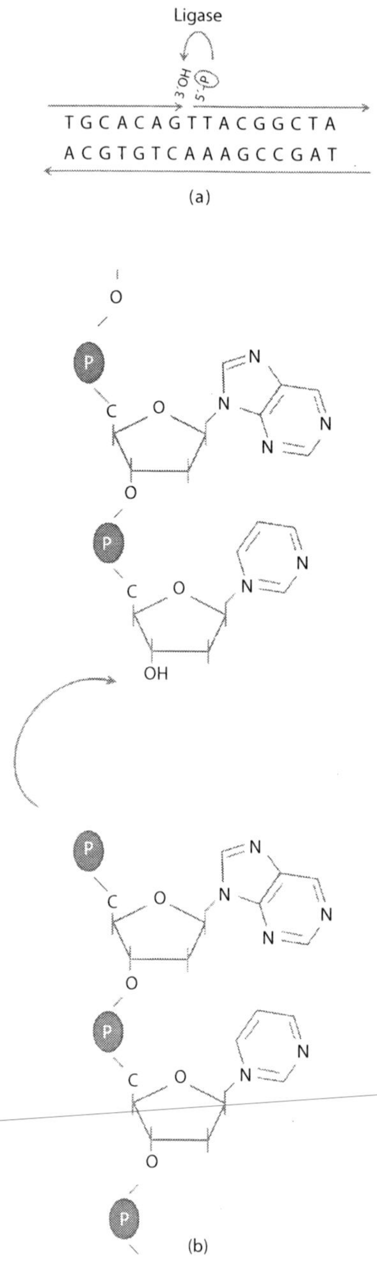

Figure 2.23 The biochemical reaction catalyzed by DNA ligases. (a) A simple depiction of a covalent bond formation between a 5′-phosphate group and a 3′-OH group by ligases. (b) A chemical diagram showing the groups involved in a ligation reaction.

source so as to carry out the catalysis of phosphodiester bond formation: The *E. coli* DNA ligase requires NADH, and the T4 DNA ligase uses ATP as the source of energy required for the ligation reaction.

T4 DNA ligase is the more commonly used ligase enzyme for genetic cloning purposes. This enzyme works very effectively to ligate DNA fragments that contain overlapping sticky ends, although it would also work on blunt-ended fragments (but a higher concentration of the enzyme and the optimal temperature of 16°C is generally used for blunt-ended ligations by this enzyme).

Ligation reactions are generally empirically optimized so as to minimize the amount of vector self-ligations or insert tandem repeats. The recommended ratio for a ligation reaction is 1 molecule of vector to 3 or 5 molecules of insert. Even then, this is an enzymatic reaction and thus it will not be 100% efficient, regardless of the vector-to-insert ratio used. Therefore, the ligation reaction will include a mixture of nonligated fragments, correctly ligated fragments, incorrectly ligated fragments, self-annealing of the plasmid vector, and sometimes even tandem ligations (Figure 2.24a).

Using two different restriction enzymes for cloning (RE1 on one end of the fragment, and RE2 on the other end of the fragment) usually circumvents one of the ligation mistakes, that is, ligation of the insert to the vector DNA in the wrong orientation (Figure 2.24b). This is particularly important if the insert DNA will be used for the expression of proteins (where the orientation will affect promoter-driven expression).

In order to avoid self-ligation or self-annealing of the vector, it is customary to treat the digested vector with phosphatases (see Section 2.3.3, "Alkaline Phosphatases") prior to ligation—this means that the vector cannot be ligated since it lacks 5′-phosphates on both strands (Figure 2.24c); there is still the problem with a vector-insert ligation reaction, however, the bacterial DNA repair mechanism will correct this "mistake" of single-strand nicks upon transformation of the bacteria.

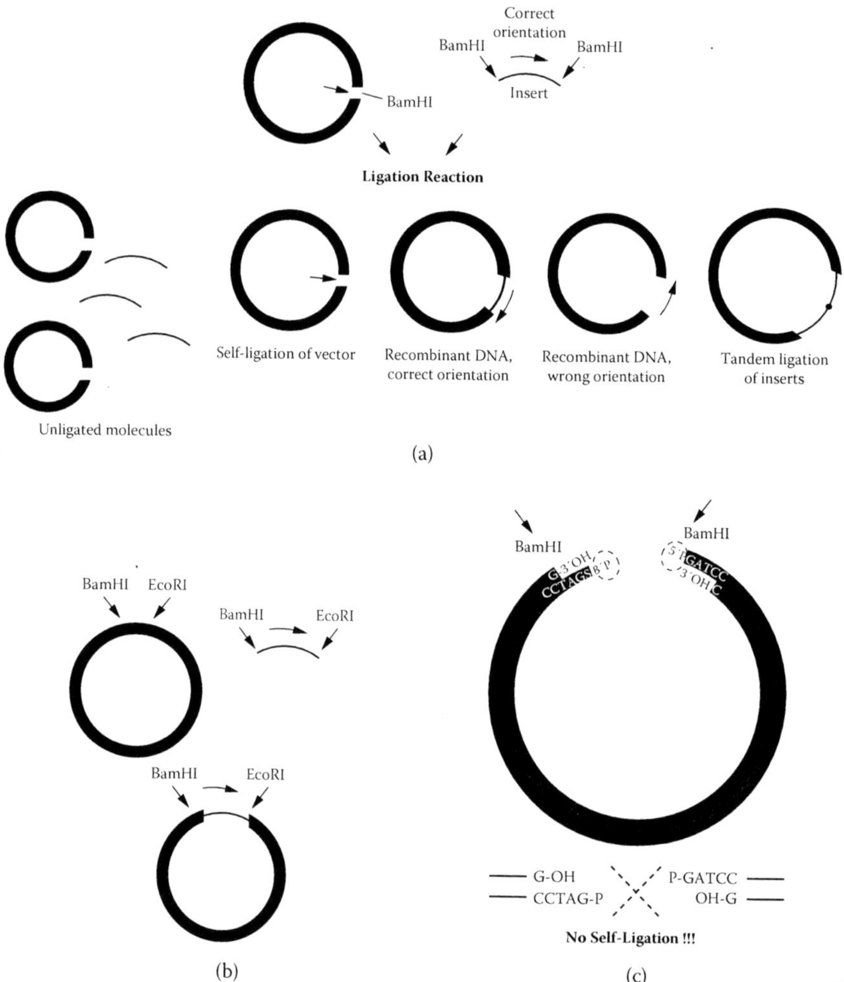

Figure 2.24 Ligation reaction. (a) Possible products of a ligation reaction, where the vector and insert are both digested with the same restriction enzyme. (b) Ligation of the insert into the vector DNA in the wrong orientation can be avoided using two different restriction enzymes. (c) Phosphatase treatment of the vector is routinely used to minimize the possibility of vector self-ligation.

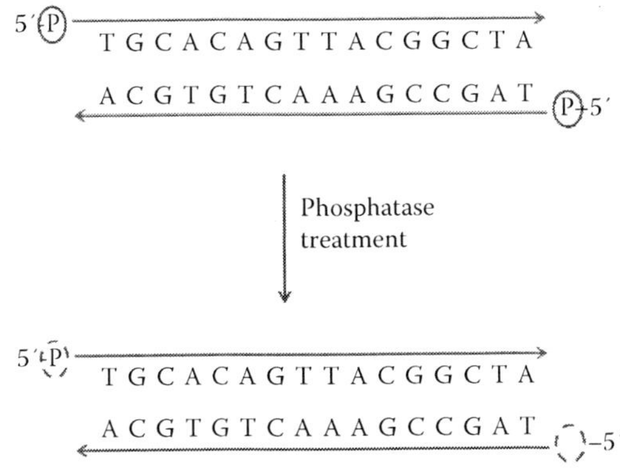

Figure 2.25 Phosphatases remove the 5′ phosphate groups from nucleic acids. APs are also used in experiments other than cloning—they are also helpful in removing the 5′ phosphate group from DNA molecules prior to radioactive labeling for synthesis of various probes.

2.3.3 *Alkaline Phosphatases*

Alkaline phosphatases (AP) modify nucleic acids by removing the 5′ phosphate groups (Figure 2.25), and are mostly active at alkaline pH. The most commonly used alkaline phosphatases in cloning laboratories are the shrimp alkaline phosphatase, derived from cold water shrimp, and calf intestinal alkaline phosphatase. These enzymes are preferred in genetic engineering experiments essentially for the relative ease of inactivation—both these enzymes can be easily inactivated by heat treatment (65°C for shrimp and 75°C for calf intestinal AP).

2.3.4 *Recombinases*

Site-specific recombination (which will be covered in Section 7.3.3 in Chapter 7) can also be employed for cloning, which in the long term is an effective way to subclone DNA sequences from one vector to many others. The typical example is the GATEWAY™ series of vectors, which is based on the bacteriophage lambda recombination mechanism, consisting of the *att* sites (*attB* site on *E. coli* and the *attP* site on the phage) and the recombinase enzyme (see phage vectors in Figure 2.14). The system also includes an entry plasmid and a donor plasmid, which contain modified *att* sites for improved efficiency. The initial cloning to the entry plasmid is generally carried out using restriction enzyme fragments or PCR products cloned through the incorporation of *attB* sites through the appropriate design of PCR primers, as can be seen in Figure A.3 in Appendix A.

2.4 Basic Principles of Cloning

Cloning means *copying*, that is, making identical copies of what one has—in the case of genetic engineering, what one tries to clone is usually the recombinant DNA, or plasmid, that one has engineered (Hartl et al. 1988; Howe 2007; Nair 2008; Primrose et al. 2006). For relatively short pieces of DNA, up to several kilo bases long, high-efficiency DNA polymerases that are commercially available can be used to amplify and copy this recombinant DNA. However, for much longer sizes, PCR amplification becomes rather impractical. Therefore, a more practical method to make genetic replicas of the recombined DNA fragments became crucial, and bacteria came to the rescue!

2.4.1 Bacterial Transformation

Once the recombinant plasmid is prepared by ligating the DNA of interest to complementary restriction sites on a relevant plasmid, one needs to efficiently transfer this recombinant plasmid into bacteria. There are many different ways by which one can achieve this—historically one of the first effective methods that researchers came up with was to transfer this DNA with phage vectors, and later this method was largely replaced by treating bacteria with calcium chloride

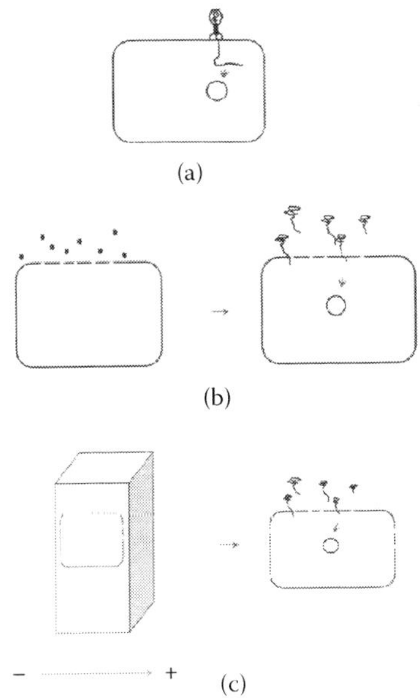

Figure 2.26 Transforming bacteria. Three of the possible methods used to transfer recombinant DNA to bacterial cells are schematized. (a) Bacteria can be transformed using phage vectors; (b) bacteria can be made competent with chemicals such as calcium chloride, which creates temporary pores in the cell wall and membrane, leaving the bacteria ready to uptake foreign DNA; or (c) bacteria can be electroporated, where the cell wall and membrane are temporarily disrupted, and foreign DNA can be transferred (the electrical field is indicated in khaki).

prior to transfer, creating transient pores in the bacterial cell wall and membrane (thus making them **competent** for the uptake of DNA), which turned out to be less costly in addition to being relatively more user-friendly (Cohen, Chang, and Hsu 1972; Hartl et al. 1988). And more recently, electroporation has become an alternative method for transformation of bacteria (Figure 2.26).

Regardless of the technique employed, transformation is never a 100% accurate process, yielding a mixture of transformed as well as nontransformed bacteria. The **transformation efficiency** (see below) of the bacteria depends on the method of transfection as well as on the size of the DNA used, among many other factors. Although in theory there is no size limitation, the transformation efficiency is still influenced by the size of the plasmid, and different methods are constantly being developed so as to enhance the transformation efficiency.

$$\text{Transformation of efficiency} = \frac{\text{\# of recombinant colonies}}{\text{\# of total competent cells used}}$$

Since transformation efficiency, no matter how close, is hardly ever 100%, the bacterial mixture will inevitably contain both nontransformants and transformants, and among the transformant bacteria some will inevitably contain nonrecombinant plasmids as well as recombinants (Figure 2.27).

2.4.2 Screening for Recombinants

There are many different ways to screen for recombinants, depending on which vector one uses for cloning. The most common method used for the screening of transformants, for instance, is antibiotic screening, although other methods such as blue-white screening and colony PCR are also available. If cloning into a phage vector, however, then a different method for choosing the recombinant phages would be needed. This section will summarize some of the basic methods of screening used for plasmid cloning.

For plasmid clonings, **antibiotic screening** is the simplest and fastest way to choose transformants. In this method, one plates the transformation mixture onto an agar plate that includes the relevant antibiotic (whichever resistance gene is present on the plasmid as a selection pressure). The nontransformants do not carry the plasmid, and therefore will not survive on the antibiotic-containing plate, whereas all the transformants (recombinant or nonrecombinant) will produce colonies due to the presence of the resistance gene

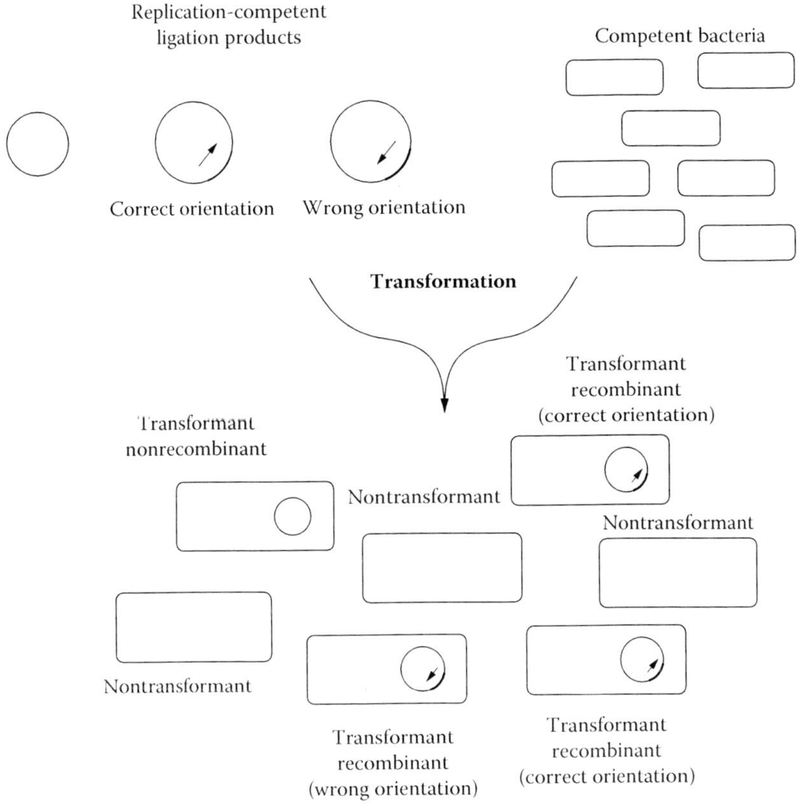

Figure 2.27 Transformation products. When competent cells are transformed with the ligation mixture, the resulting bacterial culture will contain both nontransformants, which will not grow on antibiotic selection; and transformants, which will produce colonies in an antibiotic selection. The transformants can be either nonrecombinant, containing a self-ligated vector; or recombinant, containing an insert-ligated vector with either the correct or wrong orientation of the insert.

(Figure 2.28a). **Blue-white screening**, on the other hand, is applicable to only certain vectors and makes use of the fact that certain vectors contain multiple cloning sites within their *lacZ'* coding regions. The *lacZ* gene codes for the enzyme β-galactosidase, which usually converts lactose into glucose and galactose. However, it can also modify a substrate analogue, X-gal (5-bromo-4-indoyl-B-D-galactopyranoside), from a colorless substance to a blue-colored product. This phenotype

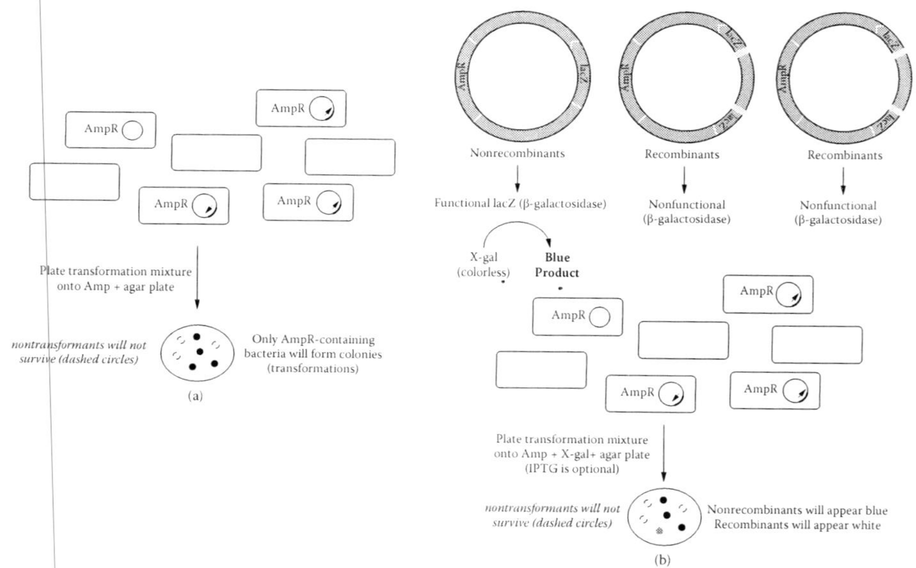

Figure 2.28 Screening for transformant colonies. (a) Antibiotic selection is a fast and economic method for screening for transformants, since nontransformant bacteria do not contain the plasmid bearing an antibiotic resistance gene and eventually die under antibiotic selection. (b) Blue-white screening can distinguish not only between transformant and nontransformant bacteria, but also between nonrecombinant versus recombinant plasmids, since nonrecombinant plasmids code for intact, functional β-galactosidase, and convert the colorless X-gal substrate into a blue product; whereas recombinant plasmids cannot produce an intact enzyme due to an interruption of the coding sequence by the insert.

could be enhanced with the inducer IPTG (isopropyl β-D-1-thiogalactopyranoside), although for many current bacterial strains used in cloning, IPTG induction is not crucial. Thus, bacteria that contain recombinant plasmids (either in the correct or incorrect orientation) will appear white due to interruption of the lacZ' coding sequence, while bacteria with nonrecombinant plasmids that code for functional lacZ' will appear blue, and nontransformants do not survive in antibiotic selection at all (Figure 2.28b).

Regardless of which screening method is used, neither is capable of distinguishing between true recombinants that contain the insert in the correct, desired orientation versus recombinants that contain the insert in the incorrect orientation. Therefore, further diagnostic screenings will be required. Two very common methods used are restriction enzyme screening (or the term preferred by this author, which is *diagnostic digestion*) and *colony PCR*.

Restriction enzyme screening, which is used to identify the correct orientation of the inserts, relies on the presence of a recognition motif for an enzyme within the MCS also within the insert, which is different than the enzyme used in cloning the insert. This is only required if the insert and vector were digested each by a single enzyme, such as EcoRI (see the hypothetical example in Figure 2.29). In such a case, use of the same enzyme on both sides is the reason for the possibility of insert ligating into the vector in either orientation to begin with (refer to Figure 2.24a for a detailed explanation, and Figure 2.24b for how to avoid this possibility). Using the same enzyme for screening will be completely uninformative, as this would simply drop out the entire insert, without giving us any clue as to which orientation the insert was cloned to begin with. However, if one can find a unique recognition motif within the insert that would generate asymmetric fragments of different sizes, that enzyme could be used to identify the orientation of the insert: in the example given in Figure 2.29, there is an XhoI site within the insert that cleaves

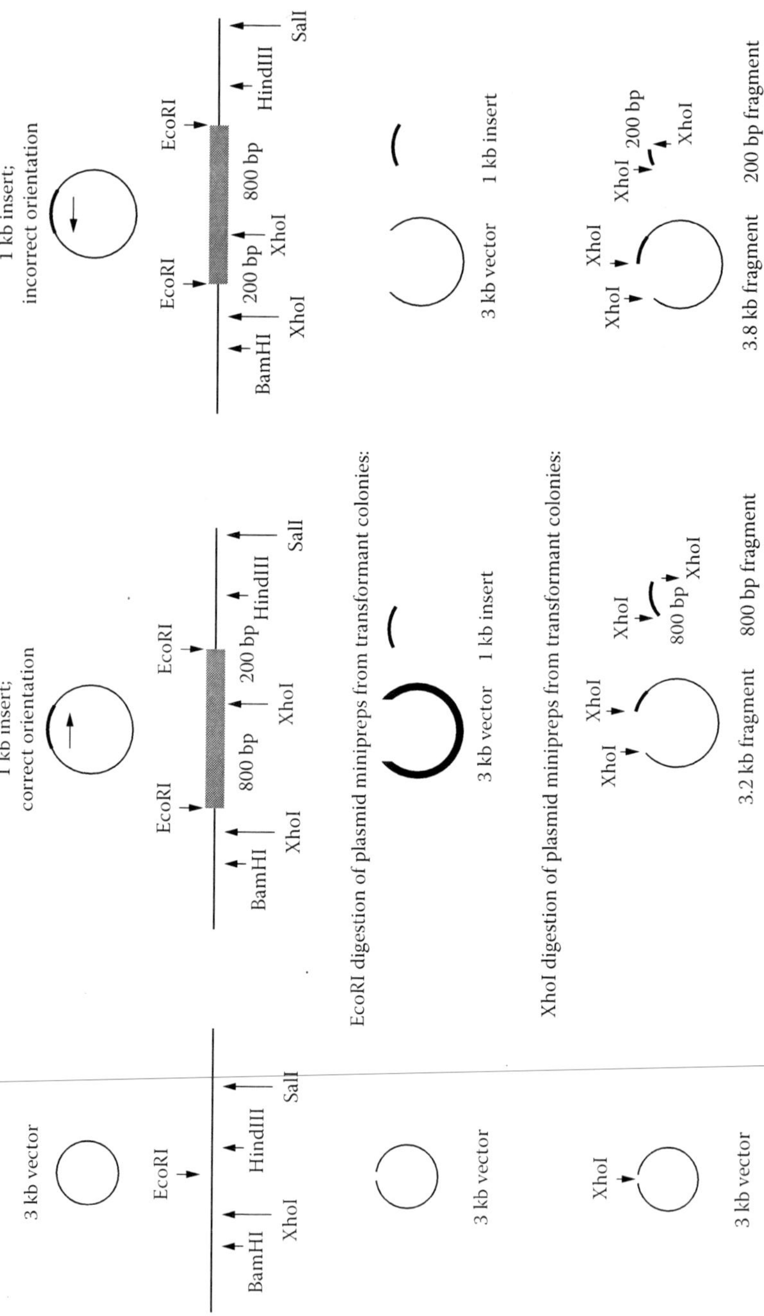

Figure 2.29 Screening for true recombinants using diagnostic restriction enzymes.

the fragment into two fragments of 800 and 200 bp. When one uses such a diagnostic restriction motif in combination with another enzyme that recognizes within the MCS, for example, the resulting digestion products would give sufficient information to determine the insert orientation (Figure 2.29).

Colony PCR is another rather rapid technique that allows for the identification of positive recombinants in a specific orientation.

2.5 Problem Session

Answers are available for the questions with an asterisk—see Appendix D.

*__Q1.__ You have the following sequence on a plasmid:

5′- ATGCTAGCA**GAATTC**TAGCTACGAT (only one of the strands is shown)

You cleave the region indicated (in bold) with the appropriate restriction enzyme, then fill in the overhangs with DNA polymerase, ligate, and reclone. You then aliquot the resulting DNA to three. In the first tube, digest the plasmid DNA with XmnI (recognition sequence GAANN/NNTTC), the second tube with Tsp509I (recognition sequence/AATT), and the third tube with AseI (recognition sequence AT/TAAT). What would be the restriction products in each tube?

*__Q2.__ A 1000 bp EcoRI restriction fragment contains a gene of interest. As a first step, construct a restriction map of the fragment using the enzymes SmaI and HindIII. The figure below shows an agarose gel of the appropriate digests. The figure shows DNA gel electrophoresis results of SmaI and HindIII digestions of an EcoRI-cut DNA fragment under investigation. Draw a restriction map of the fragment and show the distances, in base pairs, between the HindIII, SmaI, and EcoRI sites.

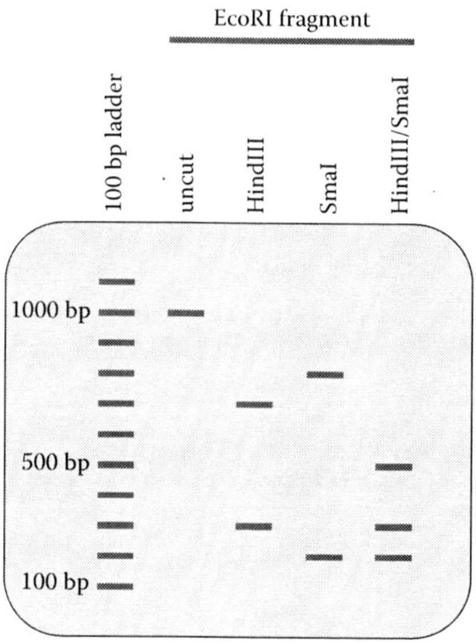

Q3. You have observed that a polymorphism linked to an X-linked disorder results in the presence or absence of an additional HindIII site within a fragment, as shown in the figure below (a). You have obtained DNA samples from a family with a history of this disorder, including their unborn baby boy. Analyzing the RFLP assay for this family shown in (b), what can you conclude about the genetic makeup of their unborn baby? The figure shows the RFLP analysis for Q3: (a) A schematic explanation of the polymorphism (the probe is shown below); (b) the results of the RFLP analysis of the pedigree shown.

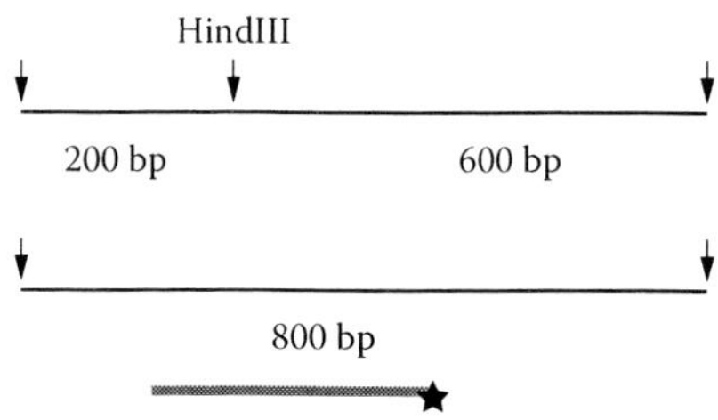

(a)

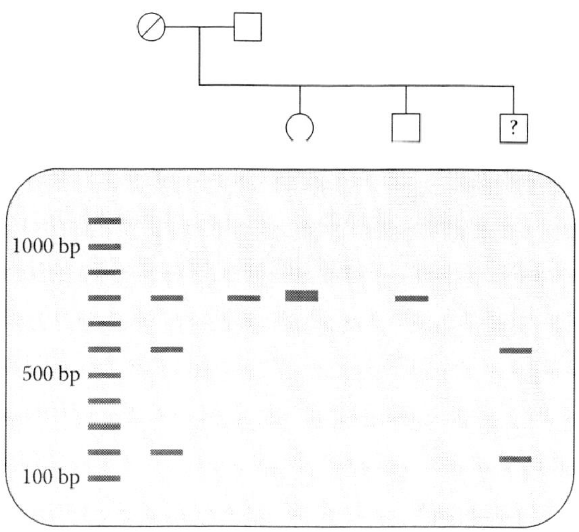

(b)

***Q4.** Find the human c-Myc coding sequence, starting from the first ATG up to the stop codon. Import this sequence to the Biology Workbench and see what you can find out about the restriction map for 6-base cutters in the first reading frame. (Hint: Use the TACG tool.)

Q5. Find the rat c-Myc sequence and align with the human sequence you have found in Q4. Which regions of this sequence have sequence identity? What other human genes share high homology to the region you identified in A3? (Hint: You need to use the BLAST tool in NCBI.)

Q6. You recently identified the sequence below from the mouse genome and you suspect that it contains part of a novel gene.

5′- GCTATGCAGCTGTAGCTAGATAGCCATGCATGCTAGCTAGTAACATAGC-3′

(a) Write the complementary strand.
(b) Find out all the reading frames and decide which one is the open reading frame.
(c) Find restriction sequences you can use for cloning, if any are present.
(d) Explain two strategies to obtain new recognition sites to this insert DNA enabling the use of specific restriction enzymes on the vector you will clone this sequence into.

***Q7.** The figure below shows the partial sequence of the pCMV-HA vector that you will use for cloning, and the insert gene that you wish to clone into this vector to create a tag fusion. How would you clone this insert into the plasmid? Explain. In the figure, the hypothetical sequences of the vector are around the MCS region (above) and the insert (below). (Note: The hypothetical insert is far too short to be a true coding sequence, but has been kept short purely for space constraints.)

HA tag

Vector ....ACC ATG TAC CCA TAC GAT GTT CCA GAT TAC GCT CTT ATG GCC ATG GAG GCC CGA ATT CGG TCG ACC GAG ATC TCT CGA GGT ACC GCG GCG GC.......

Met Tyr Pro Tyr Asp Val Pro Asp Tyr Ala

EcoRI SalI BagIII XhoI

Met Set Cys Thr Lys Gly Stop

Insert GAA TTC GGG ATC CCA ATG TCT TGC ACT AAA GGC TGA GAT CTC TCG AGC

EcoRI BamHI BglII XhoI

Q8. What would happen to your cloning strategy in Q7 if the vector you used had the HA tag at the C terminus, as in the figure below? (The hypothetical sequences of the vector to be used in this question are around the MCS region.) Explain your methodology.

HA tag

.......GCC CGA ATT CGG TCG ACC GAG ATC TCT CGA GGT ACC GCG GCC ACC TAC CCA TAC GAT GTT CCA GAT TAC GCT TAG CTT.......

EcoRI SalI BglII XhoI

Tyr Pro Tyr Asp Val Pro Asp Tyr Ala

Q9. You wish to study the sequence below in detail in your laboratory.

5′- GCTATGCAGC TGTAGCTAGATAGCCATGCATGCTAGCTAGTAACATAGC-3′

Design a cloning procedure for cloning into the pEGFP-C vector for fluorescence microscopy. (Find the vector map from the relevant databases.)

Q10. You have identified the protein sequence of a novel cell cycle-related gene:

MTEYKLVVVGAGGVGKSALTIQLIQNHFVDEYDPTIEDSYRKQVVIDGET 50
CLLDILDTAGQEEYSAMRDQYMRTGEGFLCVFAINNTKSFEDIHHYREQI 100
KRVKDSEDVPMVLVGNKCDLPSRTVDTKQAQDLARSYGIPFIETSAKTRQ 150
RVEDAFYTLVREIRQYRLKKISKEEKTPGCVKIKKCIIM 189

Design an experimental procedure to obtain the coding gene sequence for this mature protein, and to clone it into a pGEX-2T bacterial expression vector.

Q11. You are given the following plasmid construct with the gene of interest already cloned in. Your supervisor asks you to subclone the insert into another vector for expression in yeast. How would you carry out this experiment?

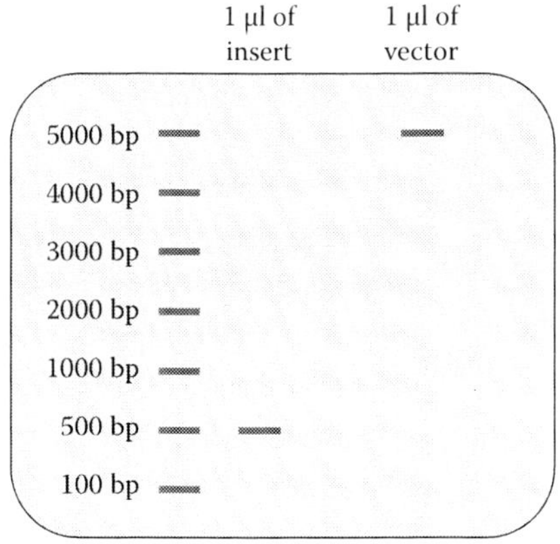

*__Q12.__ You are supposed to clone a 500 bp insert into a 5000 bp vector. (*a) To obtain a 1:5 vector:insert ratio, how much insert should you use for 0.5 μg of vector? (b) You have digested insert and vector preparations and purified the DNA molecules, ready for ligation reaction; but you do not know the absolute amounts of each DNA species. Instead, you run the preparations on a DNA gel as follows above. (The figure shows the hypothetical DNA gel electrophoresis of the insert and vector fragments to be used in cloning.)

How many μl of insert should you use for 0.1 μl of vector, if you want to achieve a 1:5 vector:insert ratio?

Q13. You are supposed to clone a 300 bp insert into a 2400 bp vector. You have digested insert and vector preparations and purified the DNA molecules, ready for ligation reaction; but you do not know the absolute amounts of each DNA species. Instead, you run the preparations on a DNA gel as in the figure below. The hypothetical DNA gel electrophoresis of the insert and vector fragments to be used in cloning are shown.

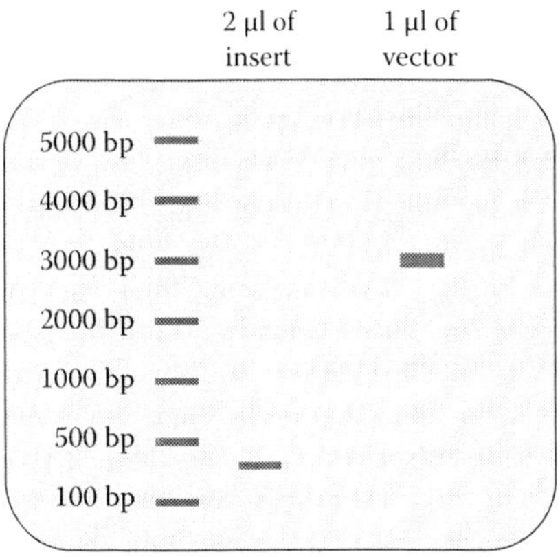

How many μl of the insert should you use for 0.1 μl of vector, if you want to achieve a 1:8 vector:insert ratio?

> The most exciting phrase to hear in science, the one that heralds new discoveries, is not "Eureka!" (I found it!) but "That's funny"...
>
> **Isaac Asimov (1920–1992)**

Chapter 3

DNA Libraries

> It is a good morning exercise for a research scientist to discard a pet hypothesis every day before breakfast. It keeps him young.
>
> **Konrad Lorenz (1903–1989)**

Introduction

A DNA library is essentially a representation of the entire DNA set (be it genomic or complementary) in a living organism, commonly in bacteria. In this chapter, we will learn about the two major types of libraries, how and why they are made, and how they are screened. A **genomic DNA library** can be defined as the representation of the entire genomic DNA of an organism, including coding and noncoding regions alike. The **cDNA library**, on the other hand, is constructed by the reverse transcription of the mRNA transcripts, and thus only represents the protein coding regions of the genome. Therefore, the two libraries are used for different purposes that will be described in detail. Based on the type of library, (a) the source of DNA, (b) the type of vector used, and (c) the experimental methodologies differ.

The basic principles of all DNA libraries involve:

1. Preparation of representative DNA fragments;
2. Preparation of the appropriate vector;
3. Cloning and transformation;
4. Selection and screening.

These steps are essentially the same as in cloning any particular DNA; and since the vectors appropriate for different types of libraries were summarized in Table 2.2 in Chapter 2, and cloning procedures described in Chapter 2, this chapter will only cover the basic principles and areas of application. (Please refer to special laboratory manuals and published articles for technical details of library construction.)

The essential difference from ordinary cloning (for sequencing or for expression, and so forth, purposes) is that the library must include up to millions of different bacteria, each of which has a different DNA fragment. Therefore, the transformation efficiency and representation/coverage become ever more important.

3.1 Genomic DNA Libraries

As mentioned before, the purpose of the study will determine what kind of library is needed, and this in turn will determine what type of vector is needed to construct that library. Once these crucial points have been identified, the genomic DNA is digested into fragments (mostly partial digestion, so as to have longer and overlapping fragments) and cloned into an appropriate vector with compatible sticky ends. These ligation products are then delivered into bacteria (either through transformation or through viral packaging and infection).

The DNA fragments, thus cloned, will not only contain coding sequences, but also introns, promoter regions, other types of regulatory regions, several repeat elements, and so on. Therefore, such libraries are useful in screening for non-coding regions of DNA. One point that should be mentioned is that since these are genomic DNA fragments, the specific tissue from where the DNA was obtained is of no significant difference as long as the tissue is from the target organism (i.e., if you want to screen for a human intron, you would have to resort to a human genomic DNA library, regardless of the tissue) (Figure 3.1).

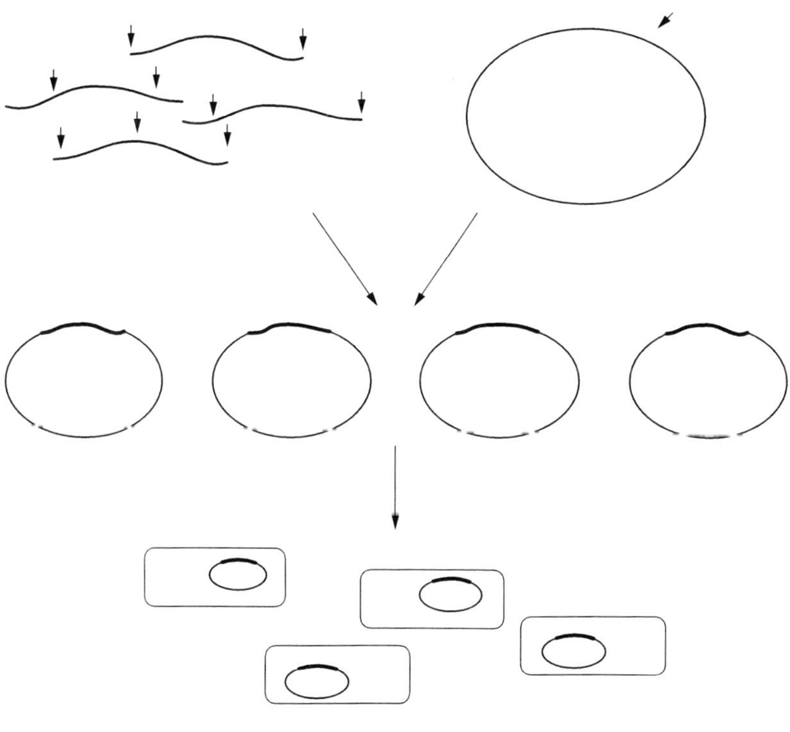

Figure 3.1 A schematic representation of the genomic DNA library principle. Essentially, almost all of the genome is targeted for fragmentation and cloning into different bacterial clones. The cloning vector is digested in this figure, and ligated with partial or full digestions of the genome; the ligation products are transformed into bacteria. Ideally, each bacterial clone should include a different genomic fragment.

3.2 cDNA Libraries

cDNA is short for **complementary DNA**, that is, DNA that has been synthesized as complementary to mature mRNA, representing only the coding regions or exons. Thus, a cDNA library refers to a representation of most, if not all, mature mRNA transcripts isolated from a cell type, or in other words, a representation of all the expressed genes in a given cell at a given time and a given condition. Consider, for instance, that insulin would not be synthesized from pancreatic β-islet cells unless there is a need for it; or a dopamine neurotransmitter synthesizing enzyme may not be present in a neuron unless there is need for it. Or, the genes expressed in a young mouse muscle will be different than those expressed in an aging mouse muscle, just like the genes expressed in normal breast tissue will be different than those expressed in benign or malignant breast tumors. Therefore, the genes that are expressed will change within the same organism, from cell type to cell type, from one developmental time to another, or from one condition to another. Much different from genomic DNA libraries, then, the source of the mRNA is of crucial importance for the specific question posed.

The first step is to identify the source from which the mRNA will be isolated. Following this, an enzyme that was first discovered in retroviruses, the so-called **reverse transcriptase** enzyme, will be used to generate DNA from these mRNA species. This is necessary for two reasons; first, RNA in general is highly susceptible to RNase degradation, RNase being a very common enzyme found even in the fingertips, therefore it becomes important to convert the RNA into a more stable DNA double helix; and second, in order to clone into a DNA-based vector one must use another double-stranded molecule that can be manipulated in a similar manner. One point to remember here is that, similar to DNA polymerases and unlike RNA polymerases, the reverse transcriptase will require a primer to start synthesizing the complementary DNA. Since the target RNA for us here is the mRNA species, it is possible to make use of the specific modifications of the mRNA, most notably the poly(A) tail. This poly(A) tail not only makes it possible for the researcher to enrich for the mature mRNA transcripts from a total RNA mixture isolated from the cells (usually through the use of oligo(dT) columns), but it also enables reverse transcription of most mRNA through the use of oligo(dT) primers that will anneal to this stretch of sequences (Figure 3.2).

The choice of vector, cloning, and transformation steps were already explained in Chapter 2, therefore, we will not go into too much detail here (suffice to say that there are a few "tricks" one will need to do to clone in the reverse transcribed fragments into a library vector), but the crucial thing to remember for the purpose of this book is that cDNA libraries thus obtained will not include any regulatory sequences such as promoters, any introns, or any other noncoding region. The cDNAs will be representative only of the genes that are expressed, that is, transcribed, modified, and translocated mature mRNA species. This also means that the cDNAs from the same organism may change from one cell type to another, from one developmental age to another, from one condition to another, unlike genomic DNA, which is essentially unchanged in almost all of the somatic cells of the same organism.

The same cDNA material cannot only be used for constructing a library, but also to directly analyze relative expression levels in different cells or cells under different conditions, in the so-called **RT-PCR (reverse transcription–polymerase chain reaction)** experiments (Section 3.4.1) and **Northern blots** (Section 3.4.2), or even microarray experiments (although some modifications to the procedure may be required—see Section 3.4.3). All three assays essentially look into the same problem: How the gene expression is affected in different cells or tissues, in the same cell at different developmental times or under different treatment conditions.

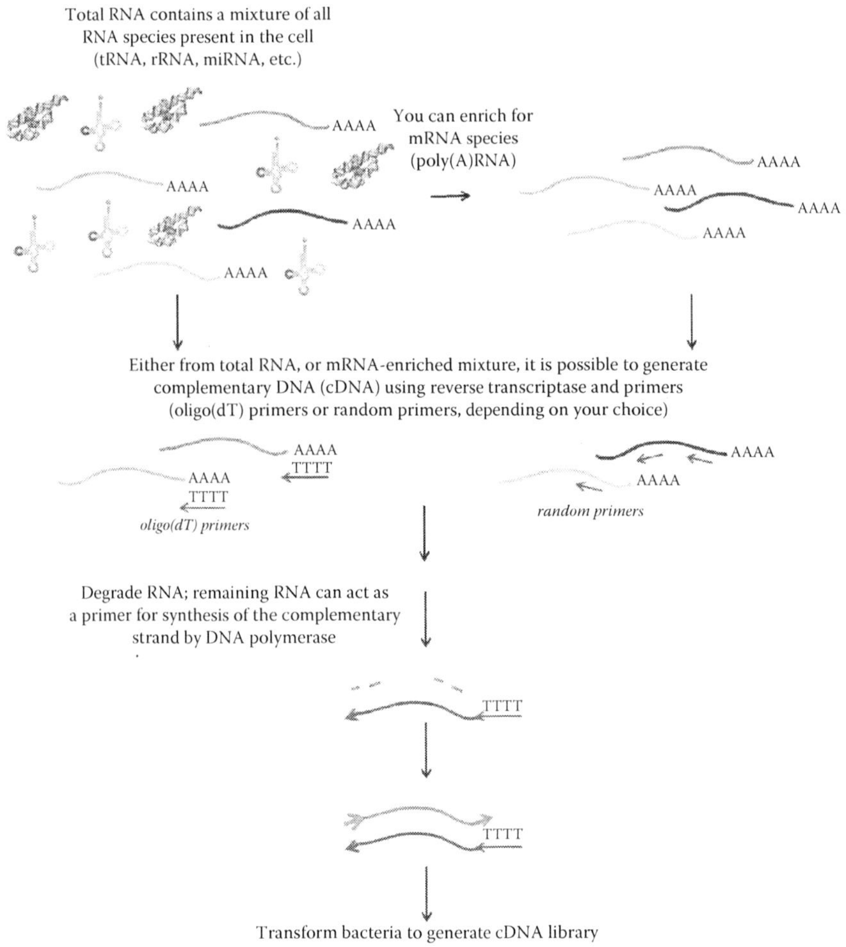

Figure 3.2 A schematic representation of cDNA synthesis and the cDNA library construction principle. Essentially, only the expressed genes (for which an mRNA transcript would be present in the cytoplasm) are targeted for cloning into different bacterial clones. However, in order to clone these expressed genes into a library vector, one must first convert the mRNA (which is highly susceptible to degradation by RNases) to cDNA (which is resistant to RNase degradation) by the use of reverse transcriptase and primers. The cDNA synthesis procedure could start from a total RNA mixture (top left), or one that is enriched for mRNA (top right).

3.3 Library Screening

Once the library has been obtained, then this library needs to be screened for a particular gene of interest, usually using a homologous DNA as a **probe**, which has to be labeled either radioactively or nonradioactively. This probe could either be the homologue of the DNA fragment from a different species (such as the coding region of a human gene used as a probe to screen a mouse cDNA library from the tissue of interest), or part of a coding region to "hunt" for a promoter region, or part of a gene from one tissue to screen for homologues in a different tissue, and so on. The basic principle behind probe-based screening is that single-stranded DNA or RNA species will hybridize to complementary sequences on the library.

There are many ways to produce a labeled probe; for example, in order to generate a radioactive DNA probe, one can first obtain the fragment to be labeled (for example, a human cDNA that had been cloned in a plasmid vector can be digested, and the cDNA fragment purified); then this fragment will be denatured (the common and easiest method is by temperature). This is then followed by extension by DNA polymerase using specific primers, but one of the nucleotides will contain a radioisotope (for instance, a mixture of dATP, dGTP, dTTP, and ^{32}P-dCTP); when the resulting *labeled* (or *radiolabeled*) DNA is denatured, the single strands which contain the radioactivity can then be used as a probe to screen for the library in question (whether genomic or cDNA).

3.4 Monitoring Transcription

When a protein-coding gene is expressed, it is first copied to an mRNA transcript. Therefore, the first level of a monitoring expression is monitoring transcription. There are a number of ways to do this—the first one described here also relies on cDNA production, namely a reverse transcription polymerase chain reaction, or RT-PCR. The second one is the traditional Northern blotting, and the third one that will be mentioned in this chapter is the microarray, or DNA chip assay.

3.4.1 RT-PCR

First, the distinction must be emphasized between the reverse transcription polymerase chain reaction that is abbreviated as RT-PCR and the real-time PCR; for the latter, Q-RT-PCR, or quantitative real-time PCR, is the preferred abbreviation so as to avoid confusion. Q-RT-PCR will be discussed later in this section.

Since RNA is relatively less stable than DNA, in other words, more amenable to degradation by RNases that are quite abundant everywhere, it is more difficult to work with the mRNA species for comparison of expression levels. Researchers therefore often refer to the cDNA that is synthesized from mRNA by reverse transcription for comparison of expression levels between different samples. As with every other experiment, an internal control is required to maintain some level of quantification: for the most part, a housekeeping gene transcript such as glyceraldehyde-3-phosphate dehydrogenase (GAPDH or G3PDH) enzyme or β-tubulin gene is used for normalization among different samples. The cDNA mix (usually the first strand is sufficient for this purpose) generated from the mRNA from different samples are then subjected to PCR amplification by primers specific for the gene of interest (Figure 3.3).

If no gene sequence is available for a favorite protein, but protein sequences are available from various species, for instance, then degenerate primers can be constructed to "fish out" the gene from the cDNA mix without the need for a library construction and screening (Figure 3.4). However, a word of caution: while library screening is long and hard, with

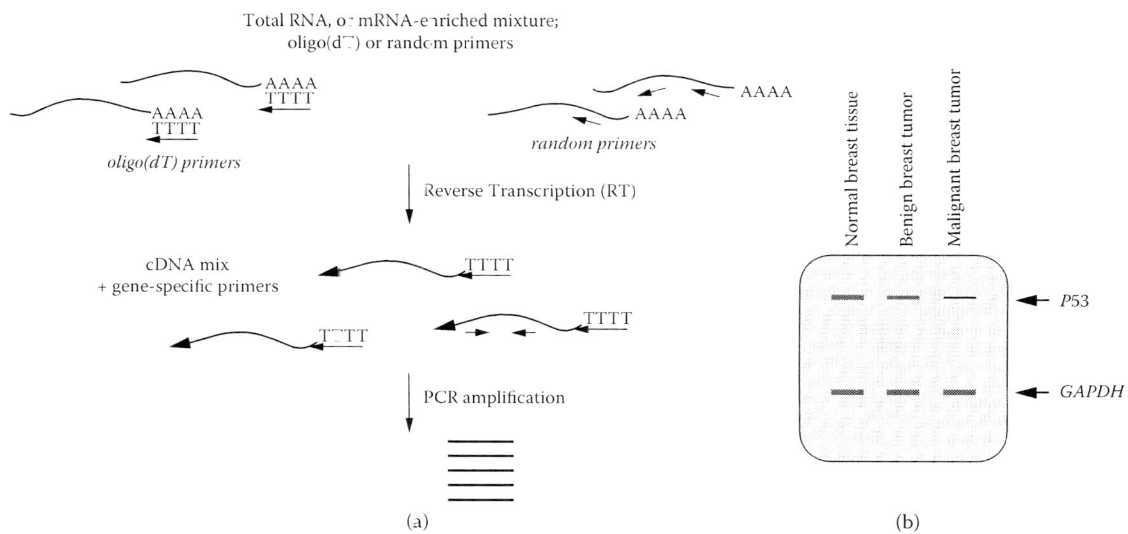

Figure 3.3 Reverse transcription polymerase chain reaction (RT-PCR). (a) A schematic summary of RT-PCR methodology. (b) A hypothetical example of an RT-PCR result on agarose gel, showing amplifications with primers specific for *GAPDH* internal control and *P53* transcript in three different tissue samples.

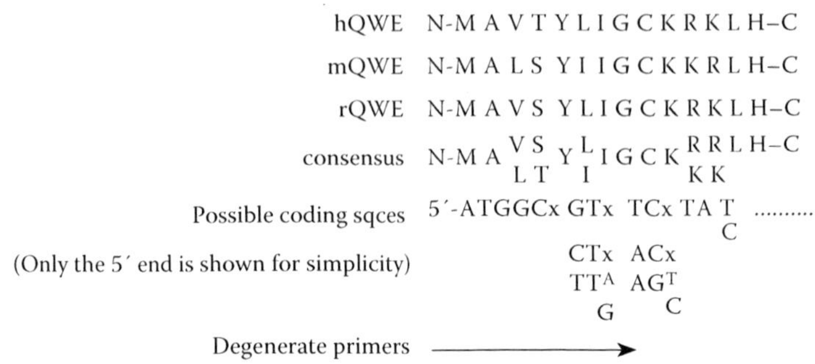

Figure 3.4 A hypothetical example of how degenerate primers can be deduced from known protein sequences through alignment. Online bioinformatic tools are also available for automated design of degenerate primers.

many possible false positives, cDNA libraries are almost certainly going to provide what you are looking for in the end, although degenerate primers will not necessarily be successful in amplifying the gene you are interested in. One should always consider these as "short-cuts," which will need quite extensive confirmations afterward, even if the PCR amplification is successful.

Quantitative real-time PCR (Q-PCR or qPCR for short) is essentially based on the same principle as PCR amplification, with the only difference being the real-time monitoring of amplification reaction. This requires a method for detection of DNA at each amplification cycle, hence the term *real time*. The detection may use a nonspecific fluorescent DNA dye such as SYBR Green, which intercalates the double-stranded DNA, hence the amount of DNA-bound SYBR Green fluorescence will increase with each cycle of PCR. Alternatively, target DNA-specific probes may be used for detection: these are usually fluorescent probes that are quenched when unbound to DNA, which only fluoresce when bound to double-stranded DNA, thus allowing for monitoring the amount of DNA in each cycle.

3.4.2 Northern Blotting

Northern blotting is a technique used for the analysis of RNA, and most commonly mRNA, among samples, and was developed by James Alwine, George Stark, and David Kemp. It was named as such after the Southern blot, developed by Edwin Southern, used for the detection of DNA. In both Southern and Northern blotting, the nucleic acids in the sample (DNA in the former and RNA in the latter) are separated in electrophoresis gel and thereafter *blotted* on a membrane, and then probed for screening (Figure 3.5).

Both in Southern and Northern blotting alike, the probes can be labeled RNA, DNA, or oligonucleotide in nature, although DNA probes are preferred because of their stability, even though the sensitivity of RNA probes is higher in Northern blots. Radioactive labeling is mostly used, although nonradioactive probes such as those using chemiluminescence can also be chosen because of their higher sensitivity and lower biohazard when compared to radioactive probes.

Essentially, probes can be synthesized in two major ways: RNA probes can be generated using *in vitro* transcription (see Section 4.2, Chapter 4), whereas DNA probes can be made by end-labeling of the cDNA fragments or PCR amplification using labeled deoxynucleotides. Short oligonucleotide probes can also be generated, in either a single- or double-stranded manner. Whichever is the method of choice, the probes must contain sufficient complementarity to the target sequence, with minimal mismatch (none, if possible).

For an *in vitro* transcription-based RNA probe, usually a bacteriophage promoter and a corresponding bacteriophage RNA polymerase is exploited, such as T7 RNA polymerase that transcribes from the T7 promoter in a pBluescript vector (see Chapter 2 for vectors, and Section 4.2 in Chapter 4 for more on *in vitro* transcription). Depending on whether the probe

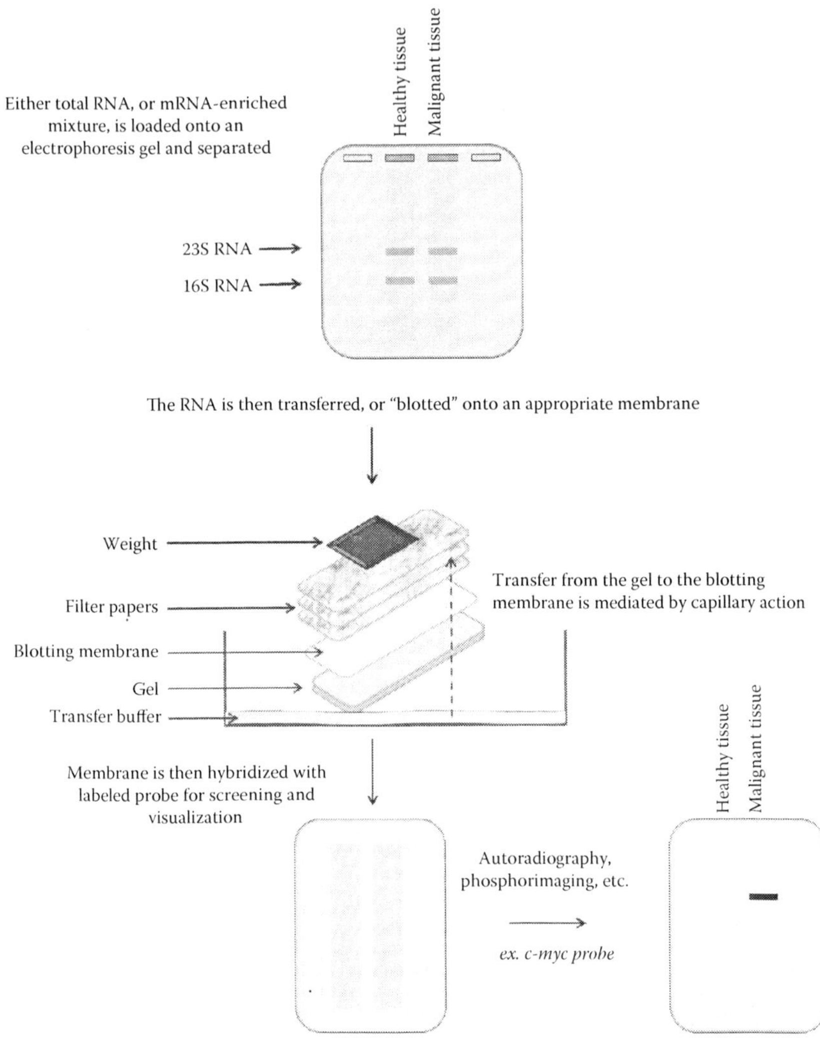

Figure 3.5 A schematized summary of Northern blotting. The RNA isolated from different samples (in this example, from healthy tissue versus a malignant tumor) are loaded onto an electrophoresis gel and separated. These bands are then transferred to a blotting membrane by capillary action. The membrane is then hybridized with a labeled probe (such as a *c-myc* probe in this hypothetical example), and the signal visualized by a number of different methods (such as autoradiography, phosphorimager, or another appropriate method).

to be synthesized should be a sense or antisense transcript, either an upstream or downstream promoter can be utilized. Usually, the *in vitro* transcription reaction contains a labeled NTP (such as DIG- or α-^{32}P-labeled UTP, along with unlabeled ATP, UTP, CTP), and the transcription product (i.e., labeled probe in Figure 3.6a) can then be used for many applications such as *in situ* hybridization, as well as for Northern blots discussed here.

For end-labeling of cDNA fragments, the DNA must first be phosphatased so as to remove the 5′-phosphate groups, and then incubated with an enzyme such as the T4 polynucleotide kinase and γ-^{32}P-labeled ATP (Figure 3.6b), followed by removal of unincorporated free ATP.

For the PCR-based preparation of a probe, the target region (probe-to-be) is amplified from template DNA by one α-^{32}P-labeled nucleotide (the rest of the nucleotides will usually be unlabeled; for example, dATP, dTTP, dGTP, and α-^{32}P-dCTP) (Figure 3.6c). If desired, nonradioactive labels such as biotin-dNTP can also be incorporated to the PCR-amplified probe.

DIG labeling is another labeling method commonly used for Southern and Northern blotting. DIG, or digoxigenin, is a steroid compound that is only found in the digitalis plants, and therefore anti-DIG antibodies are highly specific for DIG. This specificity is exploited in probe synthesis; usually a nucleotide or deoxynucleotide such as dUTP is covalently linked to DIG, and the incorporation of this labeled nucleotide into the probe can be monitored using an anti-DIG antibody conjugated with an enzyme or a fluorescent dye.

3.4.3 Nuclease Protection Assay

Nuclease protection assay, and in particular, the ribonuclease protection assay (RPA) that will be discussed here, is more sensitive than the traditional Northern blot for the detection and quantification of specific RNA species in the total RNA isolate.

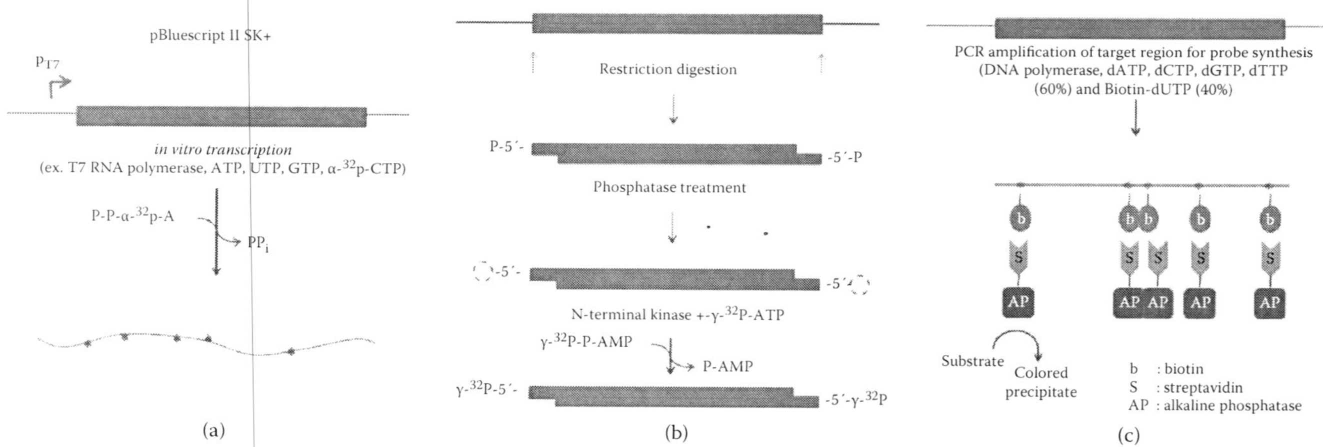

Figure 3.6 A simplified schema of probe preparation strategies. (a) Preparation of an RNA probe by *in vitro* transcription. For radioactive labeling, α-phosphate of the nucleotide that gets incorporated to the transcript should be labeled with the radioisotope. (b) End-labeling of a cDNA fragment. The γ-phosphate of ATP gets transferred to the DNA fragment. (c) PCR-based-probe generation using a biotinylated deoxynucleotide. Note that a mixture of unlabeled (dTTP) and labeled (biotin-dUTP) can be used when making the probe. The affinity of biotin to streptavidin is exploited for detection: an enzyme such as alkaline phosphatase can be conjugated to streptavidin, which then converts a colorless substrate to a colored precipitate for detection.

The principle of nuclease protection assay is the hybridization of a specific target RNA (or RNAs; multiple RNA species can be detected simultaneously, so long as the probes are of different lengths) with a target-specific antisense probe, generating a double-stranded hybrid (DNA/RNA or RNA/RNA). Any unhybridized probe or unhybridized RNA in the sample will be digested with single-stranded RNA-specific nucleases (or S1 nuclease when the probe is a DNA molecule). Thereafter, nucleases are inactivated, and the double-stranded nucleic acid hybrids are precipitated and analyzed in gel electrophoresis.

3.4.4 Microarray Analysis

In either RT-PCR or Northern blotting, one must have an idea about which gene can be expressed so that probes specific for that transcript can be prepared. However, to address a more general and unbiased question of "which genes may be expressed differently among samples," then a more general method for screening gene expression is required. Microarray or DNA chip technology is rather useful for this purpose and has become a standard for studying global gene expression profiles.

There are different microarray formats, which are usually commercially available, such as an oligonucleotide or cDNA array, cancer array, cell cycle array, toxicology array, and so forth. However, it is also possible to generate custom arrays to suit the special needs of the experimenter. Spot arrays are currently the most commonly found format, where either oligonucleotide DNA or cDNA, which is unique for each gene, is attached to a solid surface (this could be glass, nylon, silicon, or plastic). The principle of microarray is the same as reverse Northern blot, that is, nucleic acid hybridization, with the orientation reversed: the probe is unlabeled and attached to the microarray surface, and the array is screened either radioactively or, nowadays more commonly, with fluorescently labeled cDNAs from different samples (Figure 3.7).

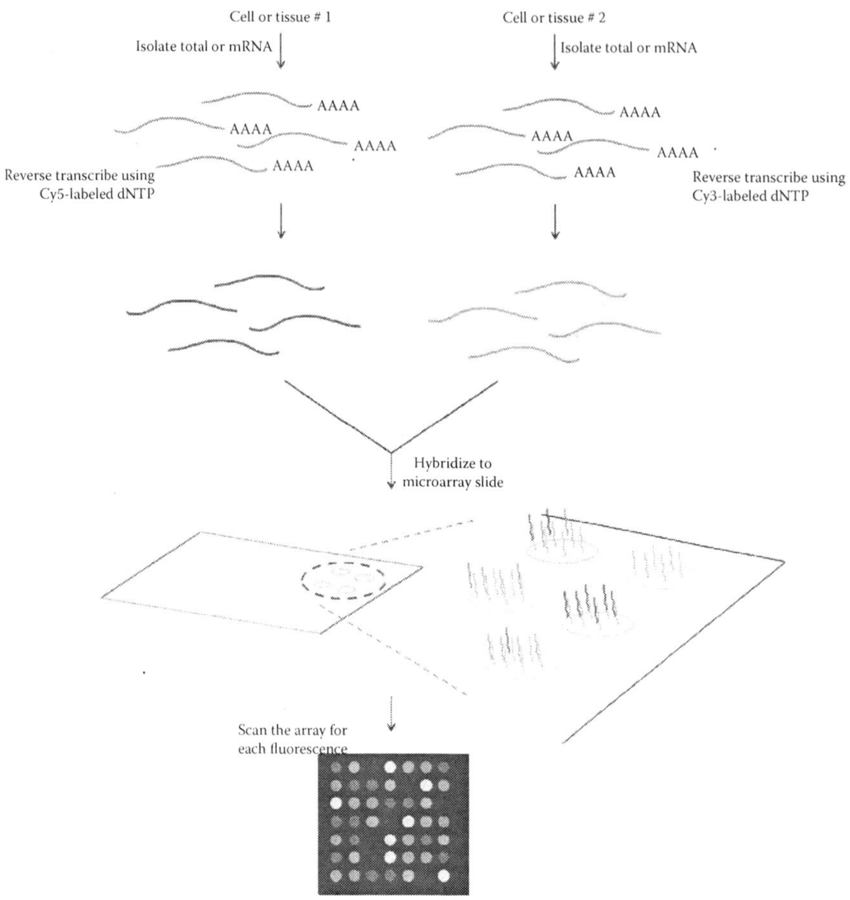

Figure 3.7 An overview of microarray analysis for gene expression profiling. Usually, RNA from two different samples are isolated and reverse transcribed using labeled deoxynucleotides (such as Cy3- and Cy5-labeled dNTPs). These labeled cDNAs are then mixed and hybridized with the microarray. Each spot on the microarray corresponds to a different gene (shown here as gray wavy lines in the zoomed-in region), and depending on which genes are transcribed in each sample (and in which amount), the labeled cDNAs will hybridize differently to each spot. When the array is scanned for each fluorescence and the information is merged, each spot will not only show a different color (different scales of red, green, or yellow—depicted in the figure as different shades of gray) but also have a different intensity measurement recorded in an Excel file or a similar file for further analysis.

The arrays may change in the number of spots that they contain (each corresponding to a different gene). The quantification, however, has the same principle: the total intensity of fluorescence from each spot is directly correlated with the number of mRNA transcripts for that gene present in the sample. These intensity measurements can be stored as Excel files, as well as displayed as colored images (Figure 3.7).

3.5 Problem Session

Answers are available for the questions with an asterisk—see Appendix D.

***Q1.** You have identified the protein sequence of a novel cell cycle-related gene:

```
MTEYKLVVVGAGGVGKSALTIQLIQNHFVDEYDPTIEDSYRKQVVIDGET    50
CLLDILDTAGQEEYSAMRDQYMRTGEGFLCVFAINNTKSFEDIHHYREQI   100
KRVKDSEDVPMVLVGNKCDLPSRTVDTKQAQDLARSYGIPFIETSAKTRQ   150
RVEDAFYTLVREIRQYRLKKISKEEKTPGCVKIKKCIIM              189
```

(a) Design an experimental procedure to obtain the promoter sequence of the gene coding for this protein. Explain your strategy.
(b) How would you study the expression pattern of this gene in tissues during development? Explain.

Q2. Gene targeting is a powerful technique that is used to analyze gene function and pathogenesis of human genetic diseases. Embryonic stem cells derived from the 129 mouse strain have been most widely used for their efficient germline colonizing ability. The use of genomic clones derived from the ES cell line is desirable for gene targeting experiments, which generates the necessity of constructing a 129 genomic library for future use.

Design an experimental strategy to generate a genomic library representing the genome of a frequently used 129 mouse ES cell line.

Q3. The WSe gene complete sequence that you have isolated and identified from human hepatocytes is as follows:

```
  1 cacttgttca atgatgtacc cccagtgtca ggcgctttgc aaacacacga tacatacggg
 61 ttgatgtttg gtcaagagag gaattaagac caggcagaca gcaggctggg atcagagaga
121 ccccatttct gtctgaaatg tctgcagaga acctggtgcc tgcctcagcc ctagctctgg
181 ggaaatgaaa gccaggctgg ggttcaaatg agggcagttt cccttcctgt gggctgctga
241 tggaacaacc ccatgacgag aaggacccag cctccaagcg gccacaccct gtgtgtctct
301 ttgtcctgcc ggcactgagg actcatccat ctgcacagct ggggcccctg ggaggagacg
361 ccatgatccc caccttcacg gctctgctct gcctcgggct gagtctgggc cccaggaccc
421 acatgcaggc agggcccctc cccaaaccca ccctctgggc tgagccaggc tctgtgatca
481 gctgggggaa ctctgtgacc atctggtgtc aggggaccct ggaggctcgg gagtaccgtc
541 tggataaaga ggaaagccca gcaccctggg acagacagaa cccactggag cccaagaaca
601 aggccagatt ctccatccca tccatgacag aggactatgc agggagatac cgctgttact
661 atcgcagccc tgtagg...........................
```

(The known coding region is underlined.)

(a) If you want to amplify the region between 241 and 481 by a PCR (polymerase chain reaction) you will need primers. Synthesize a pair of primers to amplify this region. (Keep the length of your primer between 18 and 25 bases.) How can you be sure that your primers will bind only to the Wse gene sequence?

(b) How would you study what happens to the expression of this gene in normal hepatocytes and hepatocytes that have been treated with a chemotherapeutic drug?

(c) How would you look for any mouse homologues of this gene?

(d) Having only the human WSe gene sequence, how would you search for the intronic sequences of this gene?

Q4. If you were to phosphorylate the p53 coding sequence (linear DNA) with T4 N-terminal kinase in order to generate a DNA probe for screening, which radioisotope would you use? Explain your protocol design.

Q5. You are working with a novel gene that you have identified in mouse pancreatic cells, and would like to know whether a homologue exists in humans. The first step is to clone this gene into the pBluescript (pBS) vector in order to generate a (radioactively labeled) probe. The plasmid map and sequence of the gene is shown below. Outline the steps you would take in order to clone this gene into the pBS.

5′-gag.atg.caa.ctg.caa.ctc.tgt.gtt.tat.att.tac.ctg.ttt.ctg.att.gtt.gct.tag
3′-ctc.tac.gtt.gac.gtt.gtg.aca.caa.ata.taa.atg.gac.aaa.gac.taa.caa.cga.atc

Then describe how you would generate an RNA probe for hybridization with mRNA in Northern blot assay for expression in humans.

Q6. You have identified a novel metabolic enzyme, NTR, which is expressed in aging human adipocytes, and cloned it into a pBluescript vector (shown below). You wish to obtain the regulatory region for this gene. The schematic diagram of the pBluescript-hNTR construct is shown in the figure below. How would you design this experiment?

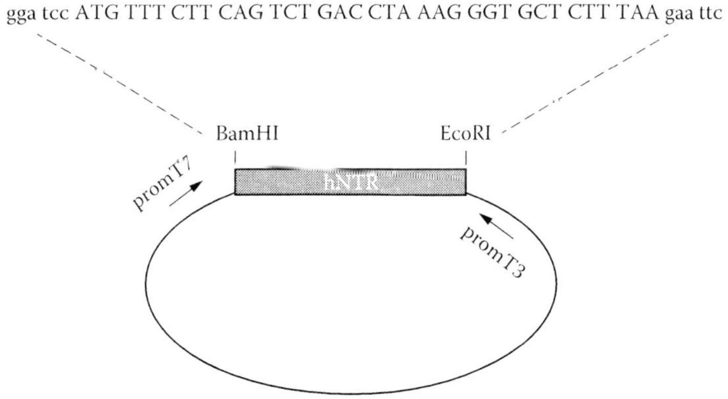

> A scientific truth does not triumph by convincing its opponents and making them see the light, but rather because its opponents eventually die and a new generation grows up that is familiar with it.
>
> **Max Planck (1858–1947)**

Chapter 4

Protein Production and Purification

There are no such things as applied sciences, only applications of science.

Louis Pasteur (1822–1895)

There are two main reasons why one would wish to purify proteins: either for analytical studies, or for preparative studies. In this book we do not concentrate on preparative studies (an advanced biochemistry book may be consulted for this), but on analytical work partly, with a particular focus on molecular cell biology applications such as cellular localization, protein–protein interactions, protein–DNA interactions, and so on. As such, the techniques described here for protein production and purification only correspond to partial (or semi-) purifications. Readers should also be reminded once again that since this is not a laboratory manual, this chapter will provide no technical details, but rather only the background and working principles behind some of the selected methods (although not every possible method can be discussed due to space constraints).

4.1 Expression Vectors and Recombinant Protein Expressions

We already discussed several expression vectors in Chapter 2, and looked into fusion proteins or tags (Section 2.2.4, "Specialist Vectors" and Section 2.2.4.3, "Expression Vectors"). The most important issue in cloning into such fusion vectors as pGEX, pCMV-HA, or similar vectors (see Figure 2.21 in Chapter 2) is to maintain the *reading frame*. Otherwise, either the frame of the gene of interest or that of the fusion partner/ tag will be shifted (see also Q7 in Chapter 2, and its answer in Appendix D, for a step-by-step explanation of in-frame fusions).

The key issue in an expression vector is, by its very nature, *expression*, which for a protein-coding gene corresponds to transcription followed by translation (Watson et al. 2008). Therefore, a good expression vector would ensure transcription of stable mRNA transcripts, which requires appropriate promoter and enhancer element(s) (Watson et al. 2008) on the vector suitable for the system or organism where transcription would take place. For example, if the gene expression will be carried out in a plant cell, a strong plant promoter must be used. If, however, one wishes more stringent control over when or where gene expression takes place, inducible promoters or tissue-specific promoters are preferred. Equally important are the transcription initiation and termination sequences: if suboptimal transcription initiation sequences (or transcriptional start sites, TSS) are used, the gene may not be efficiently expressed to desired levels. Similarly, polyadenylation sequences may be engineered to the vector so as to increase the half-life, hence the stability of the mRNA transcripts (particularly important for eukaryotic expression systems).

Since transcription is followed by translation of the protein product encoded by the gene of interest, optimal translation initiation sequences suitable for the translation system at hand must be used (Watson et al. 2008). For example, eukaryotic translation works best if the mRNA contains a Kozak's sequence prior to the start codon.

If the protein that is translated has to be secreted, targeted to an appropriate organelle, or post-translationally modified, the appropriate sequence motifs (such as localization sequences, transport sequences, phosphorylation motifs, etc.) must also be incorporated or engineered to the expression construct.

In short, a number of key design features must be considered well in advance to achieve the maximum product efficiency in your system.

4.2 *In Vitro* Transcription and Translation

The most common vector used for *in vitro* transcription and translation is the pBluescript series of phagemids (see Figure 2.18 in Chapter 2). This vector contains both T3 and T7 promoters (not shown in the map in Figure 2.18), and thus a coding region cloned within the MCS can be used to generate either sense or antisense RNA transcripts, depending on which promoter is utilized (Figure 4.1a,b).

Antisense RNA (asRNA) is a single-stranded RNA which is complementary to the mRNA sequence, and hence is transcribed using the sense strand as the template (Figure 4.1b). Although RNA is generally a short-lived and easily degraded molecule, and thus difficult to work with, antisense RNA is nevertheless a valuable tool in genetic engineering. Antisense RNA, because it hybridizes to its complementary mRNA, was shown to inhibit protein synthesis in animals as well as plants, and is thus useful for either *in situ* hybridization (which uses the hybridization property to detect presence and localization of target mRNA) or for blocking mRNA translation and effectively *knocking down* gene expression (Fire et al. 1991; Smith

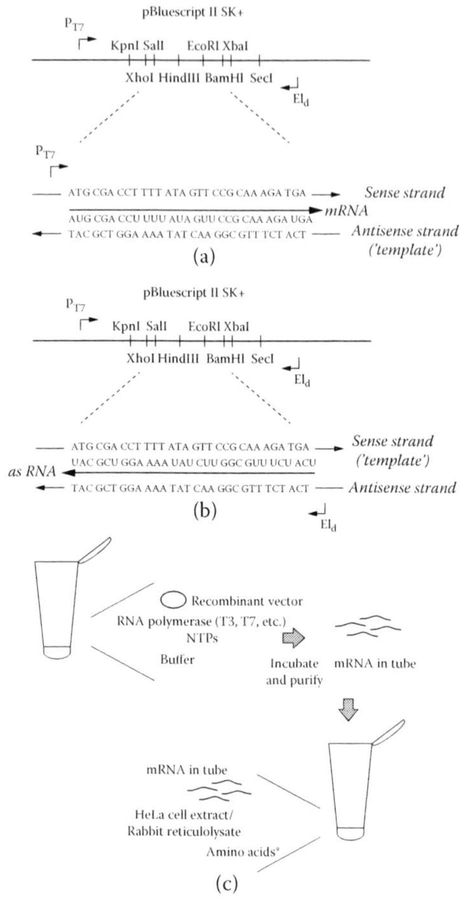

Figure 4.1 Cloning into an *in vitro* expression vector and following transcription and translation. (a) The MCS of the phagemid pBluescript SK+ is shown (see Figure 2.18 in Chapter 2 for a plasmid map). Depending on the choice of promoter, either a sense or antisense transcript may be transcribed. An example of transcribing sense mRNA from the T7 promoter is shown in (a) for antisense RNA transcription; the reverse T3 promoter should be used as shown in (b). (c) The recombinant vector is put in a reaction tube along with the appropriate phage RNA polymerase, the appropriate buffer, and the NTPs. In the second step of *in vitro* translation, the mRNA product from the first reaction and cell lysates (most commonly HeLa cell extract or rabbit reticulocyte extract-based systems) will be used along with amino acids. (*One of these amino acids is most often radioactively labeled to monitor newly synthesized proteins.)

et al. 1988). A sense-RNA probe and/or double-stranded RNA is often used as a control (which does not in theory interfere with protein synthesis); however, antisense RNA has traditionally been very tricky to work with, and results have at times been difficult to interpret, which are largely attributed to the difficulties of working with RNA. However, it was shown in the late 1990s by Fire and Mello that, in fact, double-stranded RNA is more effective than asRNA in knocking down protein synthesis, a phenomenon that they termed *RNA interference*, which is discussed in Chapter 8, Section 8.3.

4.3 Bacterial Expression of Proteins

Sometimes, proteins may be expressed inside a bacterial cell, rather than a reaction tube, such as the GST-fusion proteins (one pGEX vector map was already given in Figure 2.21 in Chapter 2 as an example). This provides an advantage over *in vitro* transcription and translation, in the sense that once the recombinant construct is transformed into bacteria and positive clones are selected, they can be stored either on the plates at 4°C, or as glycerol stocks in –80°C, and can be reutilized when needed.

The inducible nature of some operons in bacteria have been exploited in bacterial protein expression: some of the bacterial expression plasmids use, for example, the *lac* promoter of the *lac* operon, or modifications of it, to control expression from the cloned gene sequence. For induction, instead of lactose, a lactose-analog IPTG (isopropyl-β-D-thiogalactoside) is usually used. (However, remember that *lac* induction relies on the presence of a regulatable repressor in the host bacteria: if the bacteria that is chosen to express the protein in has a mutation in its *lacI* gene, which codes for the *lac* repressor, then one will not get IPTG inducibility either!)

We will return to affinity purifications (with an example from His-tag purification) in Section 4.8 of this chapter, when we cover how to create fusion protein cloning, and how to purify this fusion protein, and Chapter 6, Section 6.1, when we look into GST pull-down assays. The methods for transforming bacteria have already been discussed (Chapter 2, Section 2.4), therefore we will continue and now discuss expressions in other systems.

4.4 Expressions in Yeast

Yeast is a very popular eukaryotic model organism; it is unicellular and thus both easier to manipulate and easier to propagate; it has a cell cycle profile that is very similar to higher eukaryotes and thus is used as a model for cell cycle or cellular aging studies; its genome is deciphered (the entire 12 Mb coding for around 6000 proteins) and its metabolism is mostly known, just to name a few of the advantages. There have been quite an extensive number of tools, as well, for yeast, shuttle vectors being one of them. Shuttle vectors for yeast are those that can be used in both bacteria and yeast (hence, they would have a bacterial origin of replication, a bacterial selection, and/or a bacterial promoter, as well as a yeast replication origin, a yeast selection marker, and/or a yeast promoter) (Figure 4.2). In addition to these, YAC vectors have also been developed for other specific purposes as discussed earlier (see Figure 2.20 in Chapter 2).

Some yeast vectors are called *centromere plasmids*, since they contain the sequence from the yeast centromere, and therefore they are treated like other chromosomes during mitosis and meiosis, for stable maintenance. There are also integrative plasmids, which will integrate themselves into the yeast chromosomes and thus be maintained.

We will briefly discuss the shuttle vectors, since a similar set of vectors will be discussed later in Chapter 6, Section 6.4 for yeast two-hybrid assays. Shuttle vectors can be used both

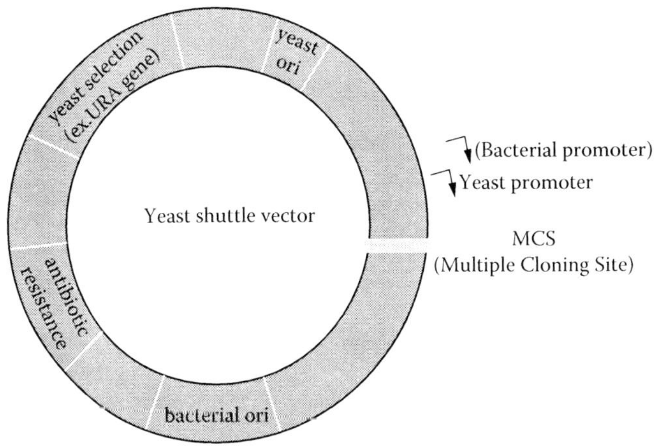

Figure 4.2 An example for a yeast shuttle vector, which could also be transformed into bacteria (for expression, if there is a bacterial promoter). The MCS is used for cloning; the yeast promoter is used for expression in yeast. Yeast origin of replication (such as the 2μ ori in 2μ plasmids is used to stably maintain the vector without integration into chromosomes) and the yeast selection marker (such as the gene of an enzyme necessary for the synthesis of an amino acid—see the text for details) is necessary for propagation and selection in yeast. On the other hand, antibiotic selection (such as ampicillin resistance) or bacterial origin of replication is used for propagation and selection in bacteria.

in yeast and bacteria, thus they also have bacterial origins of replication, and genes for antibiotic resistance for bacterial selection (Figure 4.2). Similarly, propagation and maintenance of these vectors in yeast depend on the presence of the yeast 2μ origin of replication (Figure 4.2); however, yeast selection cannot rely on antibiotic resistance, for the very simple reason that antibiotics such as penicillin or ampicillin are in fact synthesized and secreted by yeast to kill off bacteria! We must, therefore, use a different method of selection.

At this point, what comes to the rescue is the fact that eukaryotic cells can synthesize most of their amino acids within their intracellular metabolism, but they only rely on a limited number of *essential* amino acids from dietary intake.

Similarly, wild-type yeast can also take a minimum number of amino acids from the so-called yeast minimal medium, and use these as precursors for all other amino acids. However, if there is a mutation in one of the metabolic enzymes for the synthesis of one amino acid, that amino acid *has to be obtained* from external sources. Usually, host yeasts used for transformation with these vectors are mutant for this type of a synthetic enzyme, and the enzyme is provided to the yeast in the vector—in the absence of a dietary amino acid, yeast relies on the presence of the vector for the synthetic enzyme, thus the vector is stably maintained over generations (Figure 4.3).

In addition to the selection problem, yeast also presents another difference from bacteria, as well as from most higher eukaryotic systems, in that they have two different life cycle stages, a haploid phase, and a diploid phase. Therefore, they can grow both asexually, by simple mitosis or budding (depending on the yeast species) in the haploid stage, or sexually by mating of two haploids of opposite **mating types** (***MAT****a and* ***MAT***α*)*. This mating and how it is exploited in genetic engineering will be covered in Chapter 6, Section 6.4 on yeast two-hybrid assays.

4.5 Expressions in Insect Cells

Insect cells are frequently used to express high amounts of proteins from baculoviral expression vectors. Similar to yeast and other eukaryotics, insect cells are preferred for the expression of proteins toward the analysis of post-translational modifications or intracellular localizations, which cannot be addressed in a simple prokaryote that lacks such modifications or membrane-bound intracellular compartments.

The most common insect expression system involves the baculovirus-based expression systems, available from a number of different companies. In these systems, genes are usually

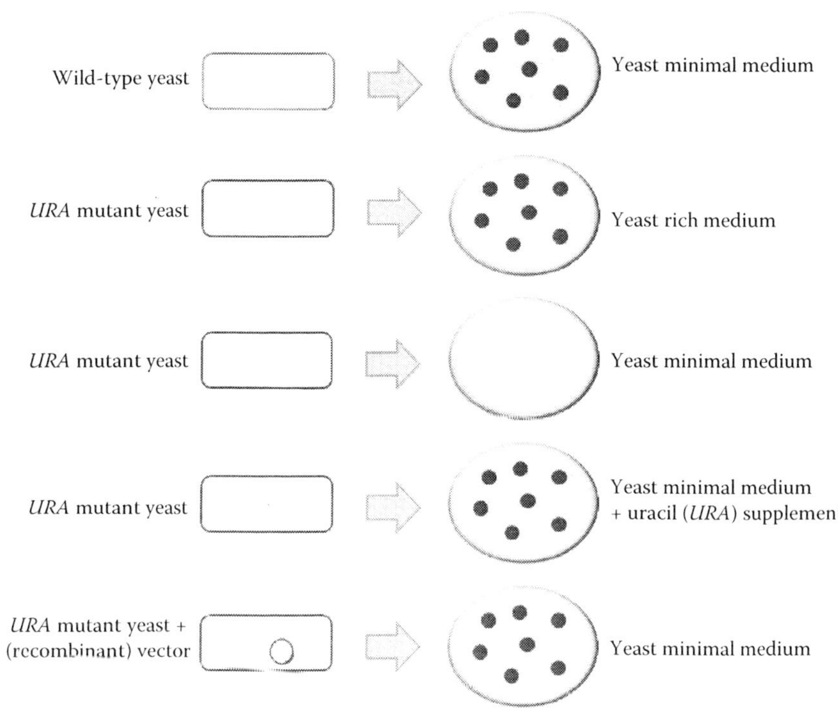

Figure 4.3 The basic principle behind selection in yeast. Wild-type yeast have all the metabolic enzymes necessary for synthesizing their entire set of amino acids, starting from basic precursors present in a minimal medium (top panel). The yeast mutant for *URA*, an enzyme necessary for the synthesis of uracil, can grow on a rich medium, which contains all the nutrients, not just the basic essentials (second panel from the top), but not on a minimal medium that lacks uracil (middle panel). The same mutant can, however, be grown on a minimal medium that is supplemented with uracil (second panel from the bottom). The only other way this mutant can grow on a minimal medium is if it has a vector that carries the synthetic enzyme for uracil, that is, wild-type *URA* (bottom panel).

cloned after a strong polyhedrin promoter of *Autographa californica* (the moth alfalfa looper) nuclear polyhedrosis virus (AcNPV); thus high levels of expression take place during late stages of infection (Figure 4.4).

A host cell, usually a derivative of Lepidopteran (the order that includes moths and butterflies) cells is used, such as Sf9,

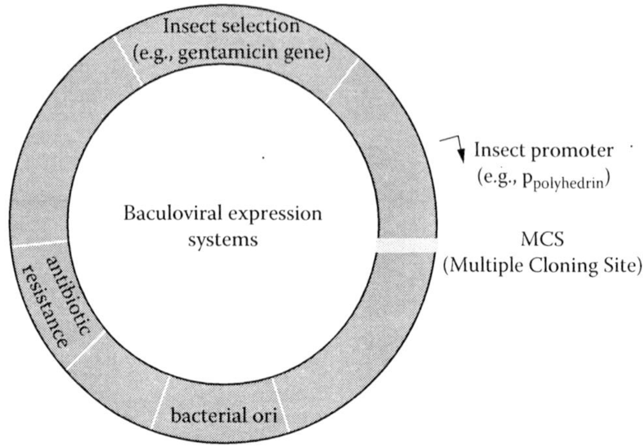

Figure 4.4 A generic baculoviral expression vector.

Sf21, High-5. Sf21 is a cell line that was originally developed from *Spodoptera frugiperda*, or the moth fall armyworm. Sf9 is a derivative of Sf21. High-5 cells (otherwise known as H5) are cloned from *Trichopulsia ni*, the moth cabbage looper. Although baculovirus expression systems can be used to infect *Drosophila* cells, there is no replication of the virus in *Drosophila*, hence it cannot infect subsequent generations. Due to this rather restricted host range of baculoviruses, they are much safer to work with when compared to other viral expression systems.

4.6 Expressions in Plant Cells

The ultrastructural features of plant cells are more similar to bacteria and yeast than mammalian cells, in the sense that DNA has to not only penetrate through a plasma membrane, but also a cell wall. Therefore, common methods for plant cell transfection include either gene guns, with which the DNA gains enough speed to penetrate through this cell wall, or stripping the cell wall and creating a protoplast that can then be electroporated, or alternatively, *Agrobacterium*-mediated DNA transfer. These methods as well as various plasmids that are available for plant expression are described in more detail in Chapter 9.

4.7 Expressions in Mammalian Cells

Mammalian cell culture is routinely used for a variety of purposes: either to study molecular mechanisms of various cellular events, or as human disease models for cancer or neurodegenerative disorders. There are a number of different culture methods, either primary cultures or cell lines, and a number of different ways to genetically manipulate these cells, be it chemical, mechanical, electrical, or viral; stable or transient; constitutive or inducible. We will discuss these in detail in Chapter 7. However, it is sufficient to emphasize that in all of these methods the gene of interest will be expressed from a promoter suitable for mammalian expression, which will be discussed in the following chapters.

4.8 Purification of Proteins

In some cases, the expressed proteins are simply analyzed by Western blots, or immunofluorescence, immunohistochemistry, or similar methods. However, in many of these cases the proteins are not available for further studies (i.e., they are either denatured for sodium dodecyl sulphate poplyacrylamide gel electrophoresis [SDS-PAGE], or fixed for immunostaining assays). In some cases, however, a native protein may be required for further biochemical or molecular analyses. In these cases, the proteins have to be purified from the cells—the extent of this purification is directly related to what we want to do with them afterward.

Biochemical and analytical-grade purification through various chromatography methods are largely covered in a variety of biochemistry textbooks, therefore we will concentrate on preparative or crude purifications (or rather, *enrichments*), based on affinity purification methods such as immunoprecipitation or GST pull-down assays.

4.8.1 Affinity Purification by Nickel Columns

Most of the crude purification kits or systems are based on noncovalent affinity interactions of one form or another, which are highly specific for the analyte and the interacting molecule, but not very robust. This method of purification relies on the presence of **tags**, or short peptide labels, on the protein being analyzed, such as the Flag tag, HA tag, or His-tag (a typical plasmid for an HA tagged clone is given in Figure 2.21 in Chapter 2). Commonly, the affinity between a tag peptide and an antibody specific for it is used for affinity purification, such as the Flag tag and the anti-Flag antibody (discussed in the following sections); however, in this section we will initially consider the interaction between His-tag and nickel.

A typical polyhistidine tag consists of at least 5 or 6 histidines at either the N- or C-terminus of a protein (Figure 4.5a) (cloning of tagged or fusion proteins was explained in Chapter 2). His-tagged proteins can thus be purified using the affinity of histidines to a metal such as nickel, as discussed in this chapter (Figure 4.5b), or to others such as cobalt resins. The idea is mostly the same as HA affinity purification: histidines present in the tag will show an affinity for the nickel in the resin or agarose *beads* and thus will be retained in the *precipitate*, whereas all the other proteins will be in the supernatant and can thus be discarded (Figure 4.5b).

After purification, the column or beads can either be used directly for further identification of interaction partners in co-immuneprecipitations, or the protein of interest may be eluted and separated (cleaved) from its tag to obtain a pure protein.

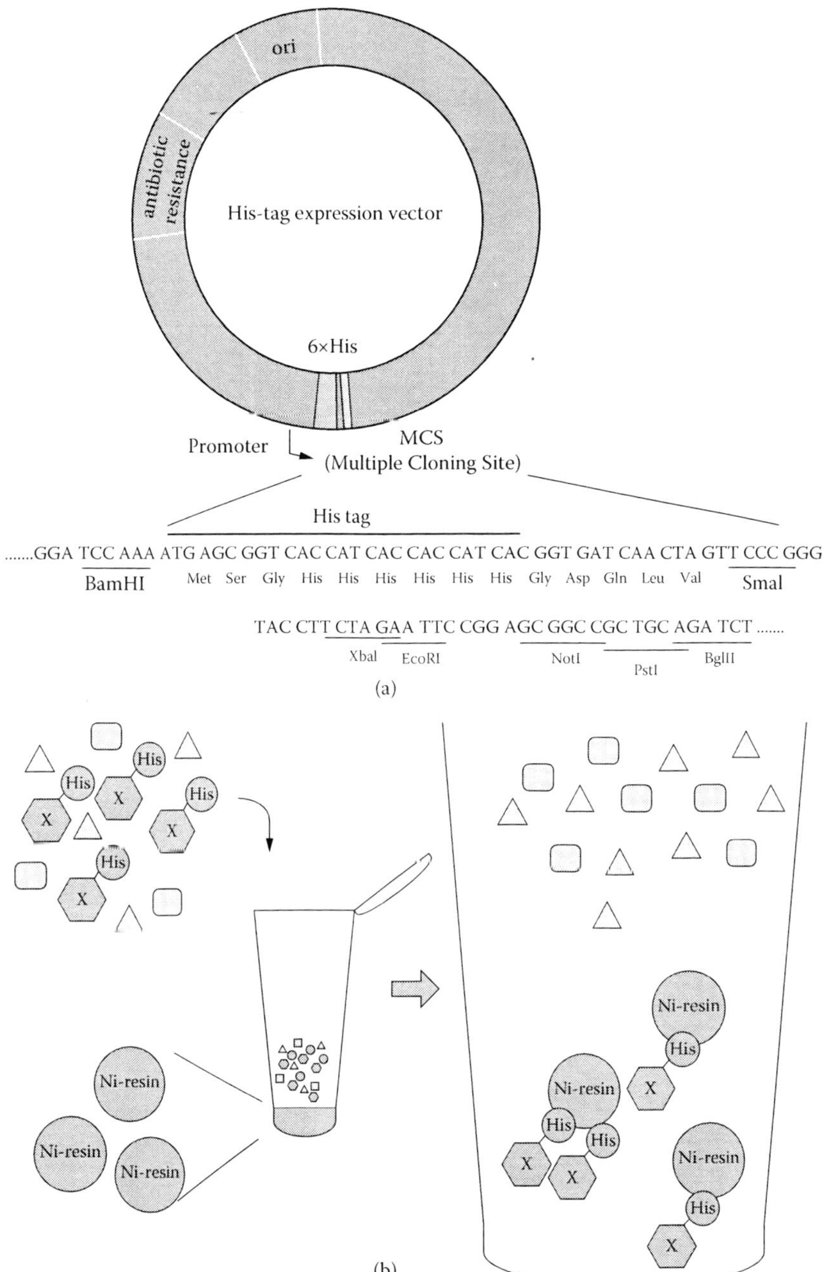

Figure 4.5 A schematic of a His-tag fusion protein purification strategy. (a) A generic His-tag expression vector. (b) Purification of His-tagged proteins using nickel affinity beads (Ni-resin). See the text for details.

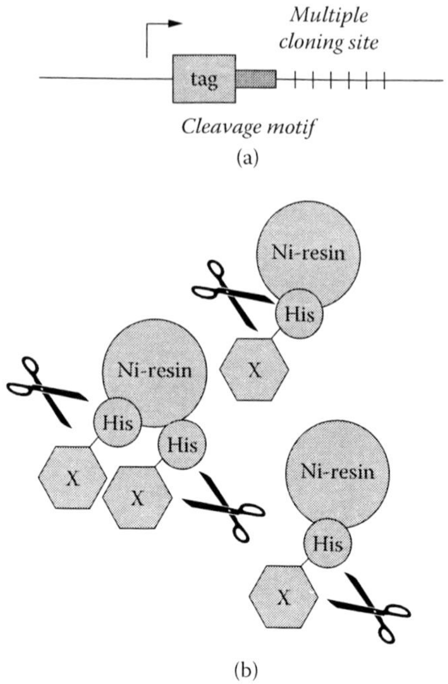

Figure 4.6 A schematic of the enzymatic cleavage of a partner protein after tag purification. (a) A general scheme for incorporation of the protease cleavage motif between a tag and the multiple cloning site. (b) Removal of protein X from its Histidine tag that is affinity-bound to a nickel resin (Ni-resin). See the text for details.

If this last option is desired, then a cleavage site for a proteolytic enzyme such as thrombin should be present in the vector used for cloning, between the tag and the multiple cloning site where the protein is to be cloned (Figure 4.6).

4.8.2 Affinity Purification Using Monoclonal and Polyclonal Antibodies

In most cases, indirect detection methods for the analysis or purification of proteins are used, and these methods will in some cases rely on the affinity of an antibody for their respective antigens.

BOX 4.1 ANTIBODIES

Antibodies are the body's defense system, produced by the immune cells. Particularly, B lymphocytes of the adaptive immunity produce antibodies specific for antigens. **Polyclonal antibody** refers to a mixture of antibodies (more commonly, immunoglobulin IgG) from different B cells circulating in the blood of the immunized animal; these antibodies are specific for a different **epitope** of the antigen. In our case this antigen is mostly the protein that we want to study (Figure B4.1). **Monoclonal antibodies**, however, are derived from a single B cell clone against a single epitope.

Polyclonal antibodies are usually obtained by immunizing an animal, typically a rabbit or sometimes a goat, by the protein that we are interested in, and after a certain period of incubation by taking a sample of the blood serum that contains immunoglobulins from various B cell clones against many different epitopes of this protein. This method is relatively easy and inexpensive to perform; however, the disadvantage is that since IgGs against many different epitopes are circulating in the blood, some cross-reactivity with similar epitopes on other proteins might occur.

Monoclonal antibodies, on the other hand, generally are more specific, since they are IgGs obtained from a single B cell clone against a single epitope, making cross-reactivity less likely. However, as the name implies, *monoclonal* antibody generation requires that there are various B cell *clones* propagated in the laboratory. In order to do this, **hybridoma** technology is used, whereby B cells from the spleen are fused with myeloma cells that are immortalized; when the clones from each

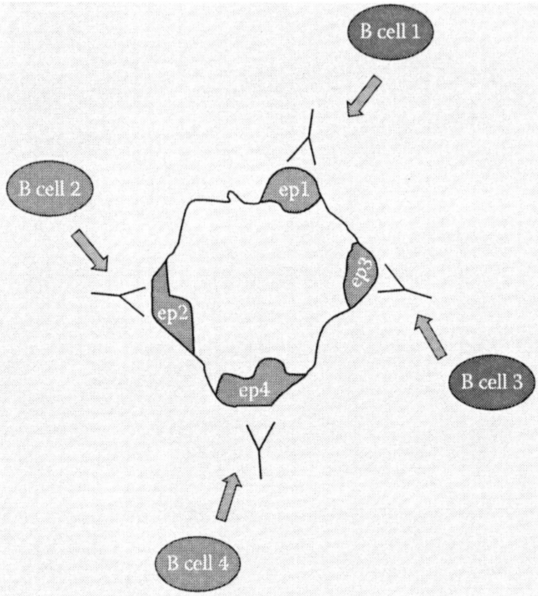

Figure B4.1 A schematic of four hypothetical epitopes of a protein antigen, and four different antibodies produced by different B cell clones against each epitope.

of the hybridoma cells are grown on multiwell plates and are screened with the antigen on an ELISA-based assay (enzyme-linked immunosorbent assay), the well that carries the specific antibody will give a reaction and the B cell clone can thus be identified. However, the length of this procedure as well as the high tech and expertise required makes this a rather expensive process.

Monoclonal antibodies are particularly useful for the purification of tagged proteins via affinity purification. For proteins that are fused to tags such as the HA or Flag, specific monoclonal antibody-bound agarose or sepharose beads or resins are available for immunoprecipitation-based purification, such as the Flag-agarose beads. The purification method is then very similar to the one discussed for His-tag nickel affinity

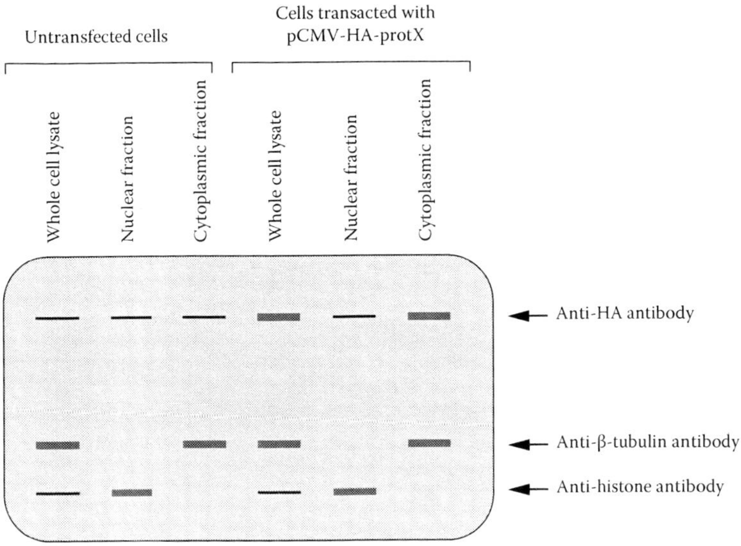

Figure 4.7 A hypothetical Western blot for monitoring the expression of protein X fused to an HA tag. The expression is studied in whole cell lysates, as well as in nuclear and cytoplasmic fractions separately. For whole cell and cytoplasmic fraction, a β-tubulin antibody is used for normalization; for nuclear fractions, a histone antibody is used. Relative expression of protein-X-HA fusion is then monitored by anti-HA primary antibody. (Secondary antibodies are not shown here for simplicity. For technical details on Western blotting please refer to Appendix B.)

purification (see Figures 4.5 and 4.6), and cloning to tagged vectors, which was discussed previously (Chapter 2).

Mono- or polyclonal antibodies can also be used for the detection of proteins expressed in a variety of ways, including immunohistochemistry, immunofluorescence, Western blots (Figure 4.7), and others, which are explained briefly in Section 4.8.3, and discussed in detail in Appendix B.

4.8.3 Monitoring Expressions in Cells

When an expression vector is used to clone a gene of interest, the purpose is obviously to study the protein product from that gene. One may wish to study one of many aspects—either the level of expression, or subcellular localization, or localization and intensity in different tissues, and so on. In order to monitor the expression of proteins, once again the specificity of antigen–antibody interaction is exploited.

The first thing one can do is to study the level of expression, in a semiquantitative manner, by a Western blot of cell lysates (see Appendix B). This is semiquantitative, since absolute quantification is not possible by antigen–antibody interaction, and a housekeeping gene such as actin or β-tubulin is used for normalization of samples, and the antibody reactivity for our protein of interest is then interpreted relative to this housekeeping gene (Figure 4.7). This method can also be modified to answer different questions, for instance, expression in a particular subcellular localization can be studied by subcellular fractionation of the cells followed by lysis and Western blotting.

Immunohistochemistry and immunofluorescence are essentially based on the same principle; however, while the former is used to study expression of proteins in tissue slices using an enzyme-conjugated secondary antibody for color reaction, the latter relies on a fluorescent dye-conjugated secondary antibody for fluorescent detection of proteins in cells (for technical details, please refer to Appendix B).

For quantitative measurements, radioactive labeling can be used. *In vitro* transcription and translation systems are perhaps the most common expression assays where radioactive labeling can be readily applied (the details of this expression system are discussed in Section 4.2 of this chapter). Basically, after transcribing from plasmid DNA using a viral RNA polymerase, buffer, and NTPs in a reaction tube, the mRNA obtained will be used to translate the protein *in vitro* with the help of a rabbit reticulolysate, a HeLa cell extract, or similar. However, instead of using a mixture of 20 amino acids, one would use a mixture of 19 unlabeled amino acids and 1

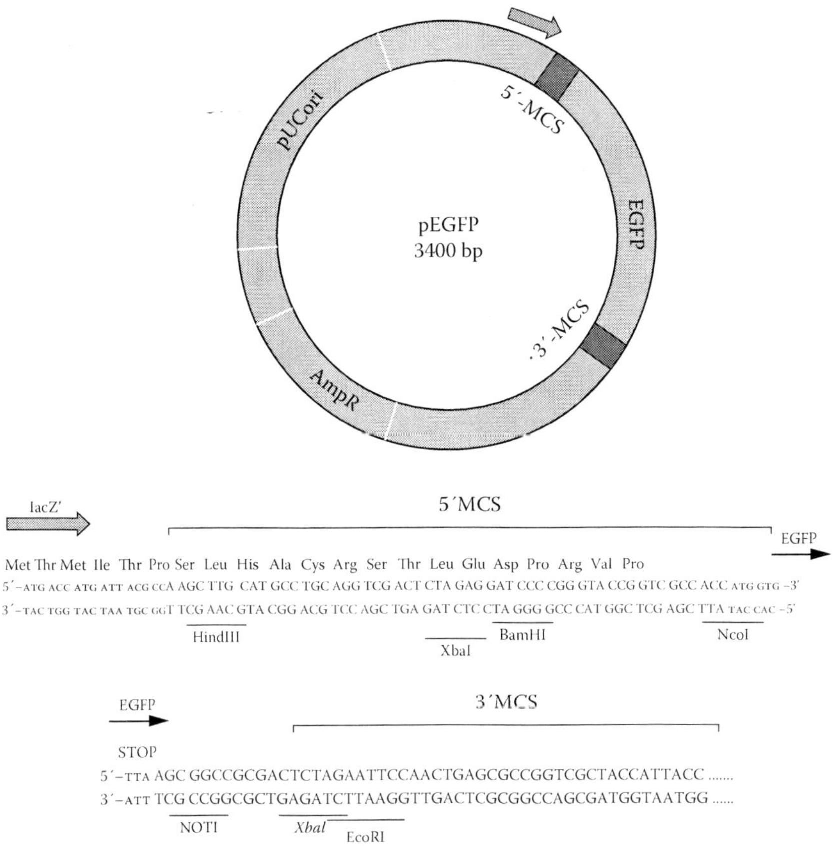

Figure 4.8 A generic map of a pEGFP vector, including the 5′- and 3′-MCS sequences. Different pEGFP variants, such as pEGFP-N1 or pEGFP-C3, and so on, exist that contain either 5′- or 3′- MCS.

radioactively labeled one, such as ^{35}S-labeled methionine. After incubation is over, the translation product is analyzed using an X-ray developer, a phosphorimager, or similar equipment to visualize the radioactivity and measure its intensity.

Pulse-chase analysis is another method used to monitor the expression of proteins, among other things, by giving a *pulse* of the labeled compound, in this case, a radioactively labeled amino acid that will be incorporated to the protein when first synthesized, and then *chase* it with the unlabeled form of the same compound. The radioactive pulse that was incorporated to the initial proteins synthesized can then be monitored for either movement or lifetime within a cell.

4.8.4 Creating Fusion Proteins: Green Fluorescent Proteins

Immunofluorescence is a useful method for monitoring the expression of proteins in cells, as well as their subcellular localizations; however, as with immunohistochemistry, immunofluorescence also relies on the use of antibodies, and as such, the fixation and permeabilization of cells so that antibodies can penetrate through the lipid bilayer. However, the cells are no longer live after this treatment, and therefore the assay can only show the presence and/or location of the protein for a snapshot in time. Given that proteins exhibit dynamism within cells, this snapshot does not present the big picture.

The discovery of fluorescent proteins has, therefore, been invaluable for the study of proteins in live cells, and for that reason was awarded the Nobel Prize in Chemistry in 2008 to its discoverers, Osamu Shimomura, Martin Chalfie, and Roger Y. Tsien, "for the discovery and development of the green fluorescent protein, GFP." (The Official Web Site of the Nobel Prize is http://www.nobelprize.org/nobel_prizes/chemistry/laureates/2008/press.html.)

GFP, first isolated from the jellyfish *Aequorea victoria,* shortly became a popular tool in molecular biology (Tsien 1998). Later, red fluorescent proteins were isolated from other species, as well as new enhanced versions of GFP such as EGFP (enhanced GFP) (a map of pEGFP is given in Figure 4.8), BFP (blue fluorescent protein), and YFP (yellow fluorescent protein) (Shaner et al. 2005). Since these proteins all have a fluorophore domain that emits fluorescence upon excitation at a certain wavelength, fixation of cells is not necessary and hence live assays can be conducted.

As shown in Chapter 2, sample Question 7 (answers can be found in Appendix D), when creating fusions to either proteins such as GFP discussed here or GST discussed elsewhere in Section 2.2.4.3, or fusions to tags such as the HA tag (also discussed in Chapter 2 and in the question named above), the crucial aspect is to (a) check the stop and start codons of coding sequences, and (b) to check that the reading frame does not shift while creating the fusion (Figure 4.9).

GFP, as well as other fluorescent proteins, are also used in bicistronic vectors for monitoring transfection simultaneously as an expression of the gene of interest. These bicistronic vectors rely on the presence of an internal ribosome entry site (IRES) between the coding sequence of the gene to be analyzed, and that of GFP (Figure 4.10). Eukaryotic cells generally translate only from the 5′-end of a transcript, unlike bacteria that can translate many coding units (cistrons) from a single transcript. IRES sequences were first described in 1988 in polioviruses as motifs that can attract ribosomes to translate from within an mRNA. This property is exploited in bicistronic vectors, where a reporter gene such as GFP is engineered after an IRES sequence, so that the gene of interest gets copied to the mRNA before the IRES and GFP sequences. Upon expression, cells that contain the reporter gene are assumed to also express the protein of interest.

4.9 Post-Translational Modifications of Proteins

When a protein-coding gene is expressed, it is first transcribed into mRNA and then translated into a polypeptide. However, this polypeptide is not necessarily functional upon synthesis and folding: a number of post-translational modifications such as glycosylation may be necessary to render the protein functional or deliver it to the correct subcellular destination via the Golgi network. Alternatively, a protein's activity may be modified by a number of modifications such as phosphorylation

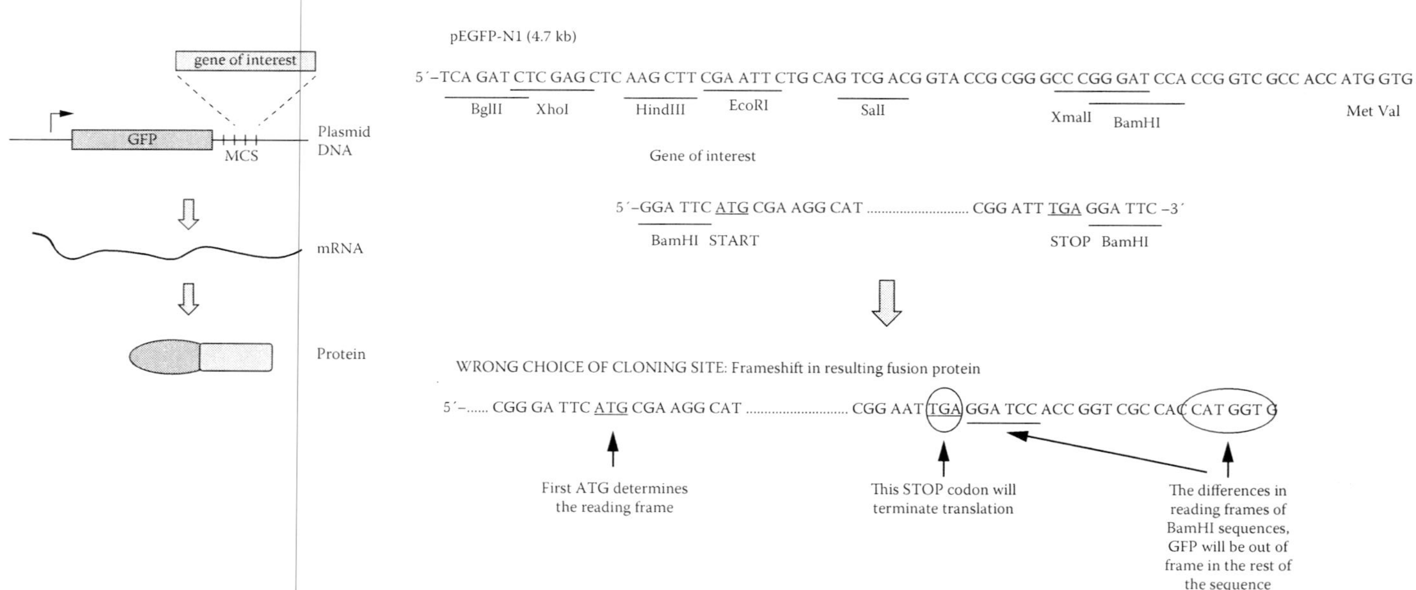

Figure 4.9 A schematic diagram showing that care must be given to start and stop codons, as well as to reading frames, while cloning fusion proteins. The left panel shows a plasmid where upon transcription and translation, GFP is at the N terminus and the protein of interest is at the C terminus of the translation product. The right panel shows a section of the pEGFP-N1 plasmid, with GFP at the C-terminus, and a hypothetical gene of interest. A wrong choice of restriction site for cloning (BamHI in this case) results in a stop codon in the middle of the fusion protein, and a frameshift in the GFP coding sequence.

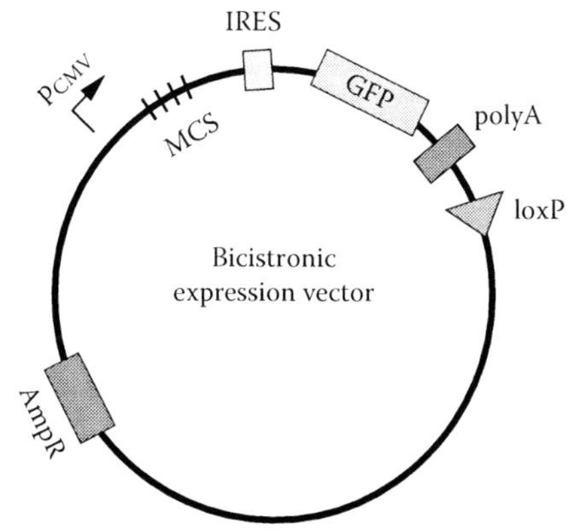

Figure 4.10 A schematic diagram of a generic bicistronic expression vector, where the gene cloned into the multiple cloning site (MCS) will be transcribed into a single mRNA along with GFP under the CMV promoter. This transcript will be used to translate two different proteins, the protein of interest, and GFP. LoxP sites are recombination sites and will be discussed in Chapter 7.

of an enzyme, sumoylation of a transcription factor, acetylation, or methylation of histone proteins. Or, the protein may be misfolded or no longer needed, and thus it gets labeled for proteolytic degradation by ubiquitination (in fact, there are many different formats of ubiquitination, each with a different meaning). Or a protein may need to be cleaved for activation (such as cleavage of a procaspase into an active caspase), or disulfide bridges may be formed for functional secreted proteins such as insulin.

Such post-translational modifications are in fact chemical modifications of a polypeptide after its synthesis and folding, and are highly regulated events. Simply monitoring the expression of a protein may not necessarily show the full range of such modifications, hence it may not necessarily tell us if the protein that is expressed is indeed active or not. 2D gel electrophoresis and other proteomic approaches, however, can also provide information about not only the level of protein expression, but also any post-translational modifications (Appendix B).

Modification-specific antibodies may be one solution to the problem: phosphor-specific or sumoylation-specific antibodies may be used in assays such as Western blots or immunohistochemistry. Site-directed mutagenesis (discussed in Chapter 5) may be used in combination to obtain information about the chemical modification of a particular amino acid and how important that residue is for the functionality of the protein.

> The vitality of thought is in adventure. Ideas won't keep. Something must be done about them.
>
> **Alfred North Whitehead (1861–1947)**

4.10 Problem Session

Answers are available for the questions with an asterisk—see Appendix D.

***Q1.**

(a) How would you design an experiment to study *phosphorylation* of a Signal Transducer and Activator of Transcription (STAT) factor upon a growth factor stimulation, either for an endogenous or for an exogenous STAT protein?

(b) How would you measure the localization of STAT in the cells upon a growth factor stimulation? Design an experiment, assuming that you only have the plasmid vectors in your lab but no cDNA for STAT.

(**Hint:** Consult the technical data sheet for antibody lists, or refer to the Internet.)

Q2. You want to study the dynamics of an enzyme QBX with its cofactor, magnesium ion, biochemically. How would you obtain a semipure QBX enzyme for further kinetic analysis?

Design a strategy, assuming that you do not have the expression plasmid for QBX in your lab yet.

Q3. You have identified the protein sequence of a novel cell cycle-related gene:

```
MTEYKLVVVGAGGVGKSALTIQLIQNHFVDEYDPTIEDSYRKQVVIDGET   50
CLLDILDTAGQEEYSAMRDQYMRTGEGFLCVFAINNTKSFEDIHHYREQI  100
KRVKDSEDVPMVLVGNKCDLPSRTVDTKQAQDLARSYGIPFIETSAKTRQ  150
RVEDAFYTLVREIRQYRLKKISKEEKTPGCVKIKKCIIM             189
```

Design an experimental procedure to obtain the *coding gene sequence* for this mature protein, and clone it into a pGEX-2T bacterial expression vector.

Q4. You have the following coding sequence of a mouse obesity-related gene in pBluescript vector, cloned between XhoI and BamHI sites (shown in boldface):

5′- **CTCGAG** GCTATGCAGCTGTAGCTAGATAGC
CATGCATGCTAGCTAGTAACATAGC **GGATCC**-3′

(a) How would you carry out *in vitro* synthesis of the protein coded by this sequence?
(b) Design a procedure for subcloning this sequence into pCMV-HA vector, and explain your strategy for purification of this synthesized protein from mammalian cells.
(c) Design a cloning procedure for cloning into pEGFP vector for fluorescence microscopy.

Q5. You have previously demonstrated in your lab that Metaphase Chromosome Protein 1 (MCP1) is involved in the early events of DNA replication. You now want to show that MCP1 associates with proteins that are required for the establishment of the pre-replication complex such as ORC2, ORC4, and MCM2. The sequence of the MCP1 protein is as follows:

N-Met-Lys-Arg-Ala-Leu-Lys-His-Val-Arg-Arg-Cys-C

(a) There are no plasmids available yet for ORC2, ORC4, or MCM2, but commercial antibodies for all three can be purchased. How would you design this experiment?
(b) You then wish to carry out immunofluorescence to study the colocalization of MCP1 with some of these proteins. How would this assay be designed?

Q6.

(a) If you were to study the expression of a p53 protein in your whole cell lysates using a Western blot, which antibody combination would you use in your experiment? (Hint: Consult the technical data sheet for antibody lists, or refer to the Internet.)
(b) If you were to study the p53 fusion to GST, how would you check whether the fusion protein is expressed in bacteria before starting your pull-down assay?
(c) p53 is, as the name implies, a 53 kDa protein. When fused to GST, how big a fusion protein would you expect? (The GST coding sequence is 633 nucleotides long.)

Q7. If the GFP fusion partner of the vector starts at vector nt 603-605 (ATG) and finishes at vector nt 1408-1410 (last Val codon), how big would the entire GFP-NXX fusion protein be in terms of kilodaltons? Your NXX coding region is 900 bp.

Q8. HUL5 is a ubiquitin ligase that increases the ubiquitylation and therefore subsequent proteasome-mediated degradation of misfolded cytosolic proteins, and its localization to cytoplasm is known to be important for its function.

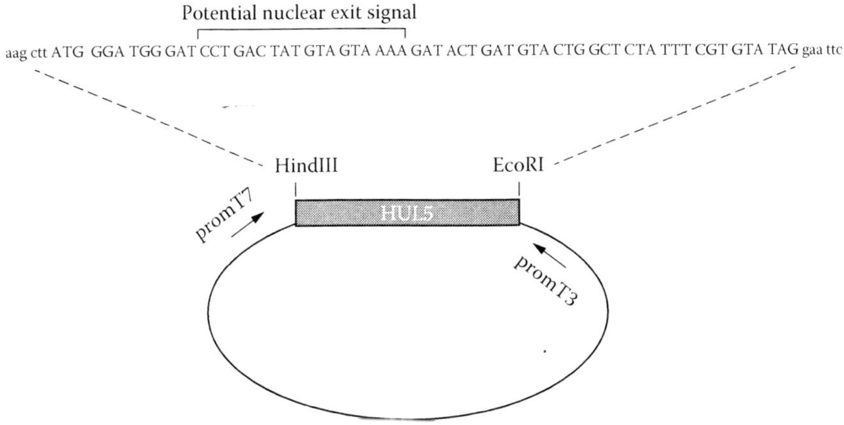

Assuming that you have identified a potential nuclear exit signal in the middle of the HUL5 coding region, how would you design an experiment to show that this region is indeed important for its localization? The HUL5 coding sequence is already cloned in the pBluescript vector as shown in the figure, which is the schematic diagram of the pBluescript-HUL5 construct. The sequence that codes for a potential nuclear exit signal is shown at the top of the diagram.

Q9. You have identified the protein sequence of a novel cell cycle-related gene:

```
MTEYKLVVVGAGGVGKSALHQLIQNHFVDEYDPTIEDSYRKQVVIDGET    50
CLLDILDTAGQEEYSAMRDQYMRTGEGFLCVFAINNTKSFEDIHHYREQI   100
KRVKDSEDVPMVLVGNKCDLPSRTVDTKQAQDLARSYGIPFIETSAKTRQ  150
RVEDAFYTLVREIRQYRLKKISKEEKTPGCVKIKKCIIM             189
```

You would like to understand what happens to the cells in the *absence* of this protein. How would you design your experiment? Describe the constructs required for this experiment and how you constructed them.

> No amount of experimentation can ever prove me right. A single experiment can prove me wrong.
>
> **Albert Einstein (1879–1955)**

Chapter 5

Mutagenesis

> Seize the moment of excited curiosity on any subject to solve your doubts. For if you let it pass, the desire may never return, and you may remain in ignorance.
>
> **William Wirt (1772–1834)**

5.1 Mutagenesis

The simplest way to understand the function of a gene is to observe what happens in its absence. Following that line of inquiry, geneticists rely on patients who display certain symptoms and inquire about the loss or deficiency of which gene(s) may be responsible for these symptoms. Molecular biologists may instead choose a more direct line of approach and mutate or delete all or part(s) of a gene in a model system (which could be bacteria, cells, animals, or plants) and observe what happens to the physiology or behavior of the organism. Mutagenesis could also be employed in order to study what the role of specific amino acids or regions in the proteins are, or alternatively, to improve or change the characteristics of the protein encoded by the gene, for research purposes as well as commercial purposes such as drug development.

In this chapter we will mainly concentrate on the methods used for such mutagenesis studies. We have also included deletion studies since these can be used to understand the role of parts or regions of the gene products. We will also discuss topics related to these methods, such as directed evolution, protein engineering, or enzyme engineering.

5.2 Deletion Studies

Deletion studies, as the name implies, relies on removal of large chunks of the gene sequence through genetic manipulations. The simplest method of creating such deletions (which can later be used for cloning purposes) is by PCR using primers that define the boundaries of the deletion. For protein coding sequences, this technique is mostly employed so as to define the function of different protein domains, such as the nuclear localization sequences (NLS), DNA-binding domains (DBD), activation domains (AD), and so on. By deleting different regions of the protein and observing the changes in its function, one can then monitor which region is responsible for that particular function.

A common practice is the deletion of regions of different sizes from both the N- and C- termini; a prior study of the gene sequence and its translation, coupled with bioinformatic analysis of putative domains, if applicable, is often necessary for intelligent design of the experiments (Figure 5.1).

The cloning of N- and C-terminal deletions is relatively straightforward, in the sense that the same restriction sites as the full-length coding region can be selected and cloned to the vector in an identical manner (Figure 5.2a). However, when internal domains within the coding sequence need to be deleted, a different strategy has to be devised. There are different ways for generating internal deletions; the one described here is one of the simplest, but may not be applicable to all the coding sequences, since restriction enzymes within the

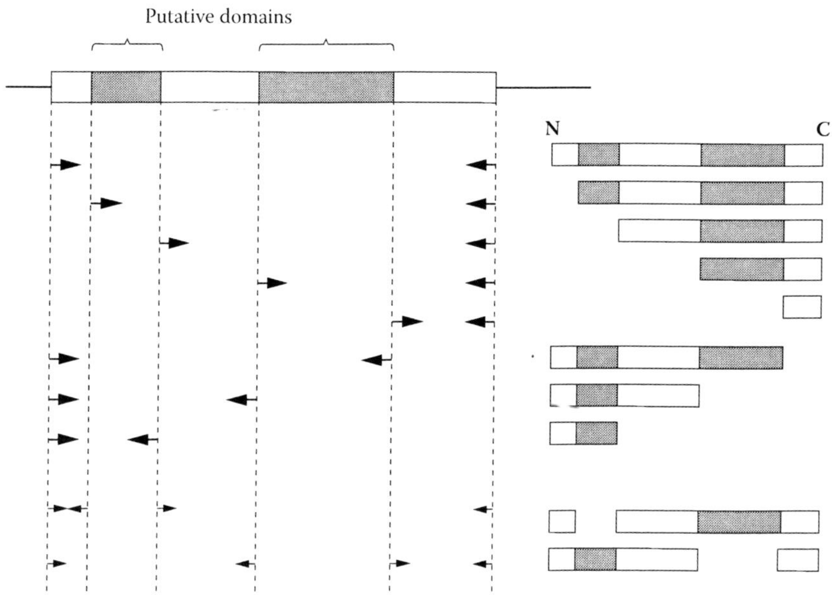

Figure 5.1 A schematic example of a typical deletion analysis of protein-coding sequences for the analysis of putative domains (active site, binding motif, interaction region, activation domain, etc.). If very little is known about the predicted domains, then a large series of N- and C-terminal deletions can be used to study the function of these domains (the forward and reverse primers that could be used for PCR amplification of these deletions are shown). Additionally or alternatively, depending on what is known and on the design of the experiment, one could also delete an internal segment (such as the putative domain) from the sequence. In that case, two sets of primers may be used, but with a well-calculated design of primers before and after the region (considering the region spans several amino acids when translated; the arrows indicate forward and reverse primer sets; note that the primers in the middle must include restriction sites that overlap and that are in-frame with the coding sequence).

vector and within the coding sequence play an important role in the design of the cloning strategy (see Chapter 2). The method described here as an example is a two-step procedure where the regions outside the target domain are amplified in two subsequent PCR reactions (Figure 5.2b; steps 1 and 2). Naturally, the design of the primers is critical in the cloning

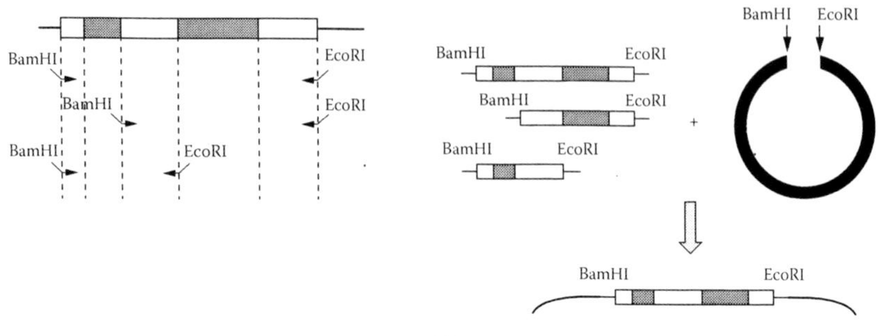

Figure 5.2 A schematic example of the cloning strategy for terminal or internal deletions. (a) For N- or C- terminal deletions, the simplest strategy would be to PCR-amplify the regions to be deleted, using appropriate forward and reverse primers. The same restriction sites as in the cloning of a full-length sequence can be used; that way, the same vector used for the cloning of the full-length sequence can be used directly. (b) For deletions of internal regions of the protein, such as specific domains spanning several amino acids when translated, many different strategies may be developed. The one shown here is just one of these possibilities and the method described here is a two-step process. In the first step, the first half of the sequence is amplified and cloned into the target vector; and in the second step, the product of the first cloning is used as the *target vector* to clone the product of the second step PCR. Therefore, the primers in the middle are designed to include the same restriction enzyme, while the outer primers are designed to have different restriction sites, if possible (see the text for details). (Note that only one of the step 1 and step 2 PCR products are shown in the cloning steps for simplicity. Enzymes shown here were chosen arbitrarily as examples.) *(Continued)*

procedure; the *internal primers* (Figure 5.2b) need to contain the same restriction site present on the vector and *not* present in the coding sequence, in addition to being in frame with both the first part and the second part of the coding sequence. Once the primers are designed appropriately, then the first half of the sequence *before* the region to be deleted is amplified, digested with appropriate enzymes, and cloned into the vector that was prepared for cloning (Figure 5.2b, step 1). After transformation and screening for positive clones, the product

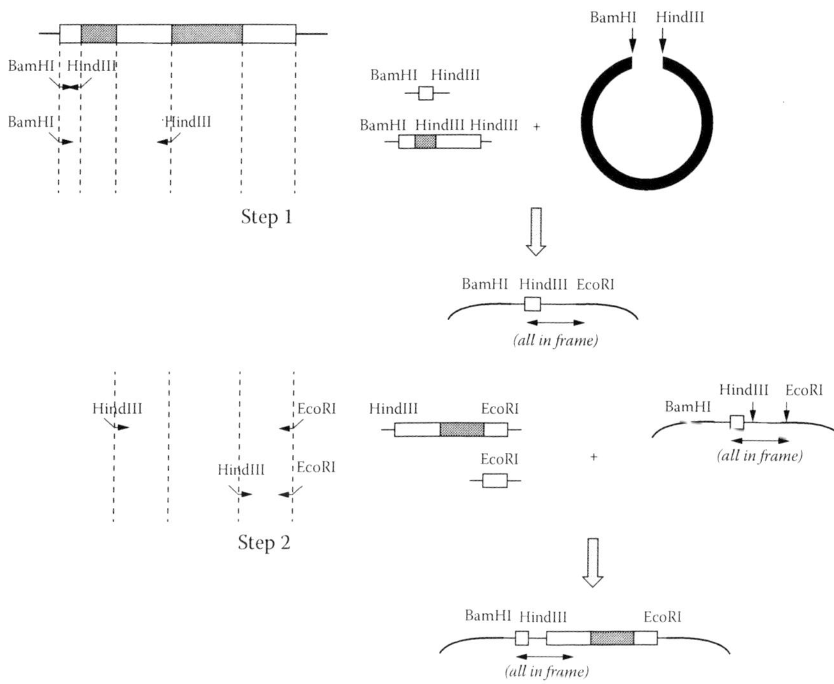

(b)

Figure 5.2 (Continued) A schematic example of the cloning strategy for terminal or internal deletions. (a) For N- or C- terminal deletions, the simplest strategy would be to PCR-amplify the regions to be deleted, using appropriate forward and reverse primers. The same restriction sites as in the cloning of a full-length sequence can be used; that way, the same vector used for the cloning of the full-length sequence can be used directly. (b) For deletions of internal regions of the protein, such as specific domains spanning several amino acids when translated, many different strategies may be developed. The one shown here is just one of these possibilities and the method described here is a two-step process. In the first step, the first half of the sequence is amplified and cloned into the target vector; and in the second step, the product of the first cloning is used as the *target vector* to clone the product of the second step PCR. Therefore, the primers in the middle are designed to include the same restriction enzyme, while the outer primers are designed to have different restriction sites, if possible (see the text for details). (Note that only one of the step 1 and step 2 PCR products are shown in the cloning steps for simplicity. Enzymes shown here were chosen arbitrarily as examples.)

BOX 5.1 PROMOTER ANALYSES

Deletion studies do not necessarily apply to proteins only. One of the routine applications is the study of regulatory regions of genes, so as to identify the essential regulatory elements. We will therefore introduce some of these assays here.

PROMOTER *BASHING* EXPERIMENTS AND REPORTER ASSAYS

These assays are commonly used to identify minimal promoter regions, enhancer elements, or other regulatory motifs important for the regulation of transcription, and are commonly combined with reporter assays so as to assess transcriptional activity of these deletion sequences. The reporter assays are described in detail in Chapter 7; therefore, we will simply refer to a generic *reporter* for the purpose of this discussion.

Promoter deletion analyses are essentially carried out in the same manner as deletion studies described in Section 5.2, referred to as *5′-deletions* and *3′-deletions* in the case of promoters. The reading frame will not be considered, as promoters are regulatory sequences and not coding sequences, and they will be cloned to reporter vectors and not expression vectors. Reporter activity can then be measured either as absolute units, relative/arbitrary units, or percentage of control, depending on the nature of the reporter used (Figure B5.1).

DNA—PROTEIN INTERACTION ASSAYS

In the analysis of regulatory sequences, the two important questions are whether specific motifs are important for transcriptional regulation (which is addressed in the reporter assays described briefly above), and which regulatory

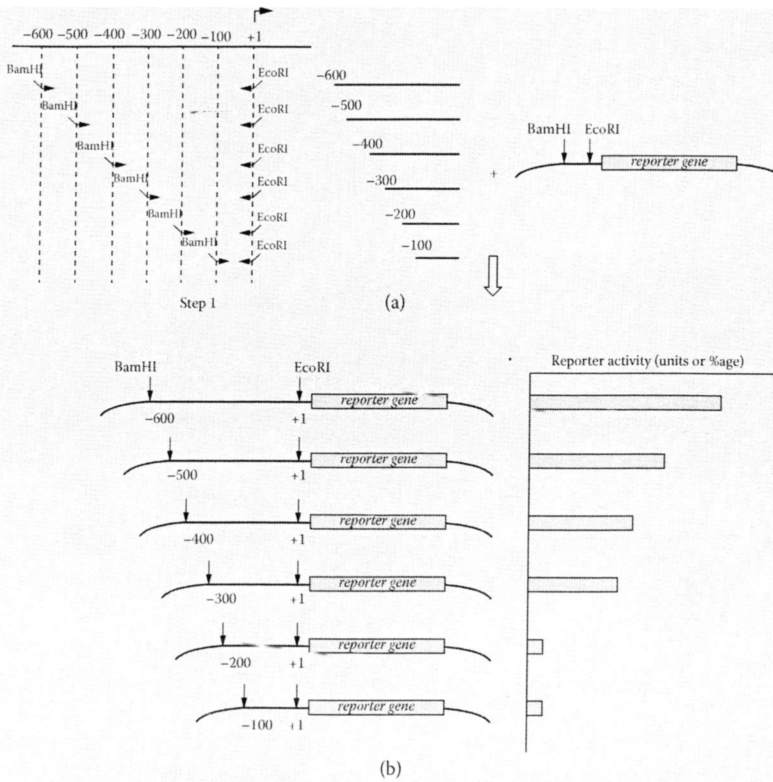

Figure B5.1 A schematic example of the cloning strategy for promoter deletion analysis and reporter assay. (a) Deletions of the 5′ end of the promoter can be constructed using different 5′ primers, and cloned into the reporter plasmid as described in Chapter 2. (b) The reporter constructs thus obtained can then be used to analyze the activity of these deletions through analysis of reporter gene activity (these will be explained in Chapter 7 in detail).

proteins bind to these motifs. This second question will be very roughly addressed in this section, which describes some of the common DNA–protein interaction assays.

ELECTROPHORETIC MOBILITY SHIFT ASSAY

The electrophoretic mobility shift assay (EMSA) is also known as the *Gel Mobility Assay, Gel Shift Assay, Bandshift*

Assay, or *Gel Retardation Assay*, among others. The basic principle of this assay is that nondenaturing gels are used, thus protein–DNA complexes are not separated or denatured and therefore both size and structure can affect migration through the gel. The most common (and the higher resolution) format makes use of radiolabeled DNA probes, although nonradioactive methods are also now available.

Essentially, this method applies to the study of DNA regions where a binding motif is known, and used as a confirmation of whether a particular transcription factor binds to that motif or not, since binding of the protein to that DNA region *retards* or *shifts* the mobility of that DNA fragment on nondenaturing gel (DNA–protein complex is much larger and heavier than DNA alone)—hence the *gel shift* (Figure B5.2).

DNA FINGERPRINTING

This method had been extensively used in the past, however, more recent and modern techniques, as well as online bioinformatic tools, have largely replaced the use of DNA fingerprinting. Still, it proves to be useful for identification of transcription binding sites in novel promoters. This method relies on the "protection" of protein-bound regions of radioactively labeled DNA fragments from DNase digestion, and thus is a straightforward way of analyzing which sequences on a given DNA segment are actually occupied (especially in combination with chemical sequencing of the same radioactively labeled fragment).

CHROMATIN IMMUNOPRECIPITATION

Chromatin immunoprecipitation (ChIP) is a method that is in fact a combination of some of the previously described

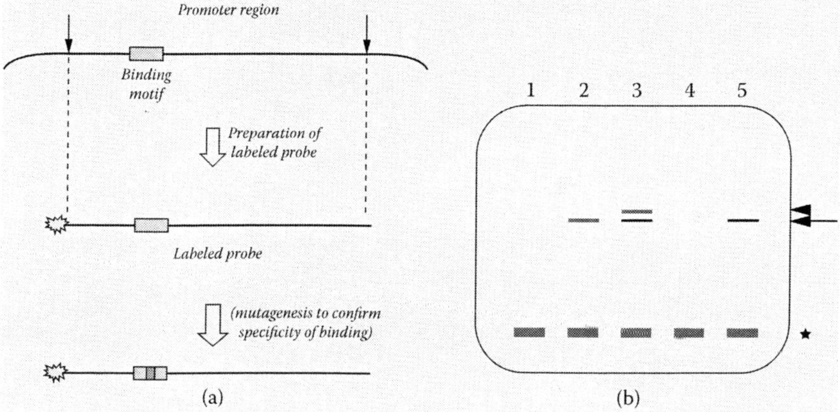

Figure B5.2 A schematic explanation of electrophoretic mobility shift assay (EMSA). (a) The DNA fragment of interest (usually a part of the promoter containing the binding motif) is used to prepare a labeled (radioactive or nonradioactive) probe (this DNA fragment could also contain a mutation introduced to disrupt the binding motif, to which the transcription factor in question is no longer expected to bind). (b) A diagram of a hypothetical bandshift assay, where the labeled probe or its mutated counterpart schematized in (a) is incubated together with either a purified transcription factor or a nuclear extract in the presence or absence of an antibody and then subjected to nondenaturing gel electrophoresis: 1, labeled probe alone; 2, labeled probe incubated with either purified protein or a nuclear extract; 3, labeled probe incubated with purified protein/nuclear extract in the presence of a specific antibody against the transcription factor of interest; 4, labeled mutagenized probe alone; 5, labeled mutagenized probe with the protein or nuclear extract. The *star* shows the probes without any protein, running fast in the gel; the *arrow* shows the retarded or shifted probe DNA/protein complex; and the *arrowhead* indicates the probe DNA/protein/antibody complex, further retarded in the electrophoretic field. (The radioactive EMSA could further be used to calculate binding constants, if the bands on the gel are quantified in experiments designed for binding affinity calculations.)

methods, along with immunoprecipitation, which will be covered in Chapter 6. What one in principle wishes to address is whether a protein is actually bound to a putative binding site *in vivo*. To address this question, cells are usually treated with a cross-linking reagent, which cross-links to any molecule in close proximity (protein–protein, protein–DNA, DNA–DNA, protein–RNA, and so on) (see Figure B5.3a,b). One then digests this complex with micrococcal DNase or else uses sonication to shear away any unprotected DNA (DNA regions that are bound by proteins will be *protected* from such digestion) (Figure B5.3c).

If one then employs immunoprecipitation with a specific antibody, the protein of interest would be precipitated with beads, along with any interaction partner, be it protein or DNA (Figure B5.3d) (nonspecific IgG can be used as a negative control, so as to help confirm specificity of the antibody-IP results). Upon precipitation, and washing away of any unwanted (unbound) protein complexes, cross-linking is reversed and any protein is digested using protease treatment. The DNA that has been precipitated along with protein-antibody complexes are purified, and subjected to PCR amplification using primers specific to the region of interest (Figure B5.3e,f). After gel electrophoresis of both the *input* (i.e., the sample before immunoprecipitation (IP), in other words, the cell lysate—see Chapter 6) and the IP, the amplification products are analyzed. A positive signal with the antibody-IP (along with a negative signal in IgG-IP) confirms specific binding of the protein in question to the target DNA region.

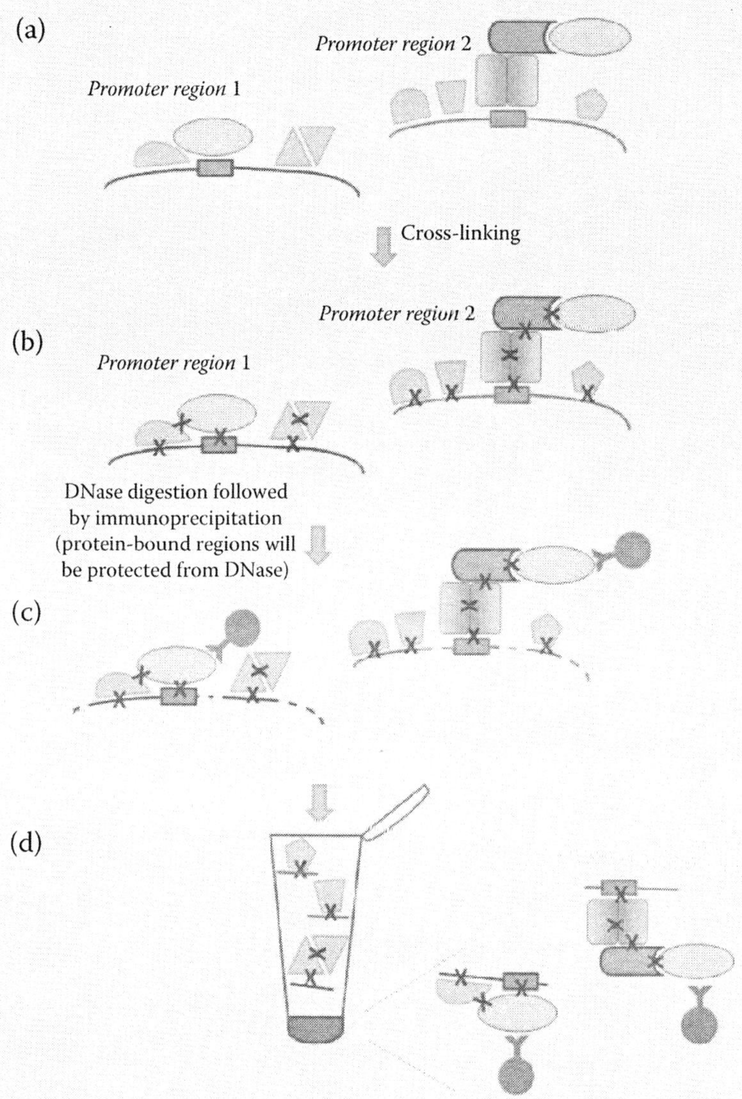

Figure B5.3 A schematic explanation of chromatin immunoprecipitation. Steps (a) through (g) are explained in the text in great detail. ***(Continued)***

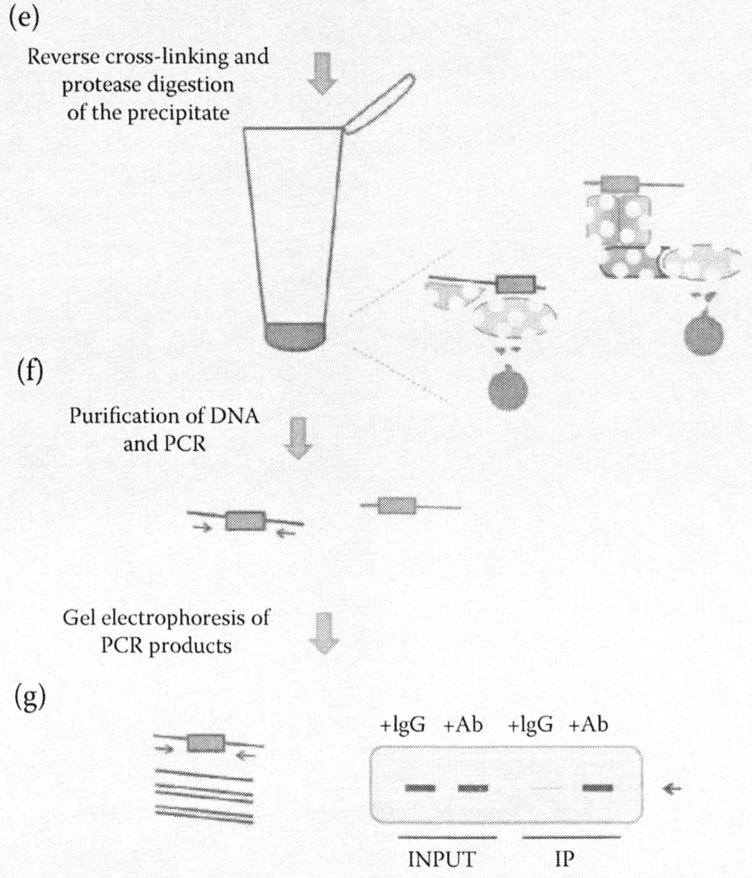

Figure B5.3 (Continued) A schematic explanation of chromatin immunoprecipitation. Steps (a) through (g) are explained in the text in great detail.

of the first step can now be used as the vector for the second step: the second half of the sequence *after* the region to be deleted can now be amplified and digested with the appropriate enzymes. Likewise, the product of the first step cloning is digested with the same enzymes, and cloning is carried out as described previously (see Chapter 2) (Figure 5.2b, step 2).

5.3 Site-Directed Mutagenesis

Site-directed mutagenesis, as the name implies, refers to the generation of mutations at specific regions of the gene sequence, with a specific purpose (such as changing a phosphorylation site, catalytic activity, ligand binding property, etc.). There are various methods by which such site-directed mutations can be created, but for space constraints we will only concentrate on a simple PCR-based strategy.

In this method, a pair of primers is designed specifically for the region of interest, introducing the desired mutation through mismatched nucleotides in the primer; for example, a potential phosphorylation motif, threonine, is to be mutated into an alanine in the sequence (Figure 5.3). In order to achieve this, it is sufficient to change the wild-type coding sequence for threonine (ACG in Figure 5.3) to one of the codons for alanine (GCG in Figure 5.3). Two such overlapping primers are then designed, along with the two outermost primers (forward and reverse primers, shown in Figure 5.3 to have two different restriction motifs for further cloning purposes). Each primer is then used in combination with either the forward or reverse primer to amplify the sequence in two separate regions, while introducing the intended mutation in the target site (hence, *site* directed). The products of this first step of PCR are then used in a second PCR, initially as primers for each other briefly: when the double-stranded PCR products are denatured and annealed in these first few cycles, each single strand may cross-anneal to the single strand from the other PCR product (Figure 5.3), and only one of these cross-annealed pairs will be productive. This second PCR step will be continued with the addition of the forward and reverse primers, thereby generating the complete mutated sequence (Figure 5.3).

There are also other methods for site-directed mutagenesis, which rely on the PCR amplification of the entire plasmid with a pair of mutation-introducing primers. Such whole

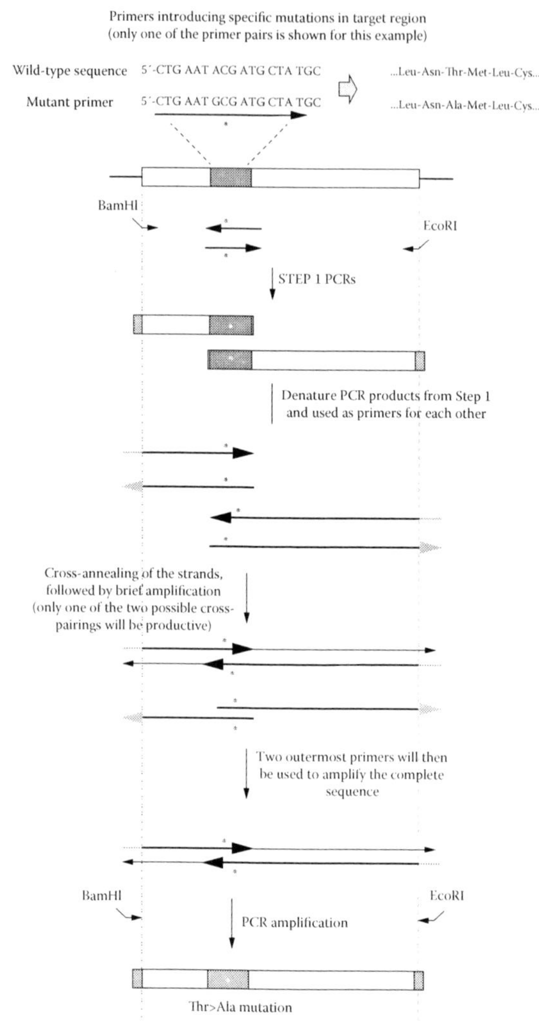

Figure 5.3 A schematic diagram of a PCR-based site-directed mutagenesis strategy. In this particular example, a potential phosphorylation motif, threonine, is to be mutated into an alanine (T > A mutation). A mutant primer pair is designed to contain a mismatched nucleotide that creates this conversion, and teamed up with forward and reverse primers (shown in this example to contain two different restriction enzymes for cloning purposes) in the first step of PCR. The products from this reaction are then used to cross-anneal in a brief PCR reaction, after which the complete mutant sequence is further amplified with the help of forward and reverse primers.

plasmid amplification strategies are relatively easy and efficient, as well as commercially available. In some of the whole-plasmid sequencing approaches, where the parental plasmid DNA is wild-type, and the newly synthesized copy is mutated as above, a different trick is employed to distinguish the wild-type parental strand from the mutated daughter strand (otherwise, only half of the transformed bacteria will have the mutant plasmid). This strategy relies on the fact that a restriction enzyme called *DpnI* will only cleave the motif GATC if the adenine is methylated (i.e., $G^{m}ATC$), but not cleave the unmethylated sequence. The plasmid is initially propagated in a *dam+* bacterial strain; *dam* being a DNA methylase that transfers a methyl group to the adenine in the same motif (i.e., GATC). Thus, the parental plasmid will have methylated GATC, whereas the newly synthesized DNA using PCR amplification (*in vitro*) will not contain this methylation, which can be used afterward to digest away the parental plasmid by DpnI, leaving only the newly synthesized plasmids for transformation.

5.4 Random Mutagenesis

Mutations have originally been introduced to genes (or rather, organisms) by either irradiation or chemical mutagenesis (for example, in Morgan's *Drosophila* experiments). While such chemical or physical mutagenesis methods are quite robust for introducing mutations to genes, they are highly hazardous in nature and thus not thoroughly desirable. Nowadays, a number of other methods are being used to generate random mutations in genes, including error-prone PCR, degenerate oligonucleotide primers, or mutant bacterial strains (such as the *Eschericia coli* strain XL1red that is defective in its DNA repair proteins so that the mutation rate is around 5000-fold higher than the wild-type strain) among many others. Error-prone PCR simply relies on the fact that *Taq* DNA polymerase or similar polymerases can incorporate wrong nucleotides in the presence of high levels of Mg^{2+} along with Mn^{2+}. Degenerate primers, on the other hand, would generate site-directed random mutants (Figure 5.4). This strategy is essentially based on the same principle as site-directed mutagenesis, with the simple difference being that rather than mutating a specific nucleotide (or a set of nucleotides) into other known nucleotides (such as A > G in Figure 5.3, introducing a specific Thr > Ala mutation), a degenerate sequence is created in certain residues (N, for any nucleotide, in Figure 5.4b) to introduce relatively random mutations at specific residues.

Such random mutagenesis products can be used to generate a mutant library so as to study structure-function relations of proteins or enzymes, improve catalytic properties of enzymes, or design proteins with desired properties. However, because a rather large collection of mutants is generated, high-throughput screening methods need to be designed for selection of these desired properties.

5.5 Directed Evolution, Protein Engineering, and Enzyme Engineering

A directed evolution experiment is briefly intended to artificially *evolve* a protein or an enzyme optimized for the desired properties (i.e., *directed* toward a certain goal). There are different means to achieve this end, albeit the most common ones include random mutagenesis, and mostly error-prone PCR, or the use of a highly mutagenic strain. The gene of interest is thus mutated randomly, creating a large mutant library from which to select the desired trait under a selection pressure, such as heavy metal tolerance, salt tolerance, higher affinity to substrate, and so forth. Once the positives are selected, the mutants are then amplified and their DNA is isolated so that the mutation resulting in the desired trait can be sequenced and identified. This represents the first *round* or *generation* of directed evolution: in many cases, multiple

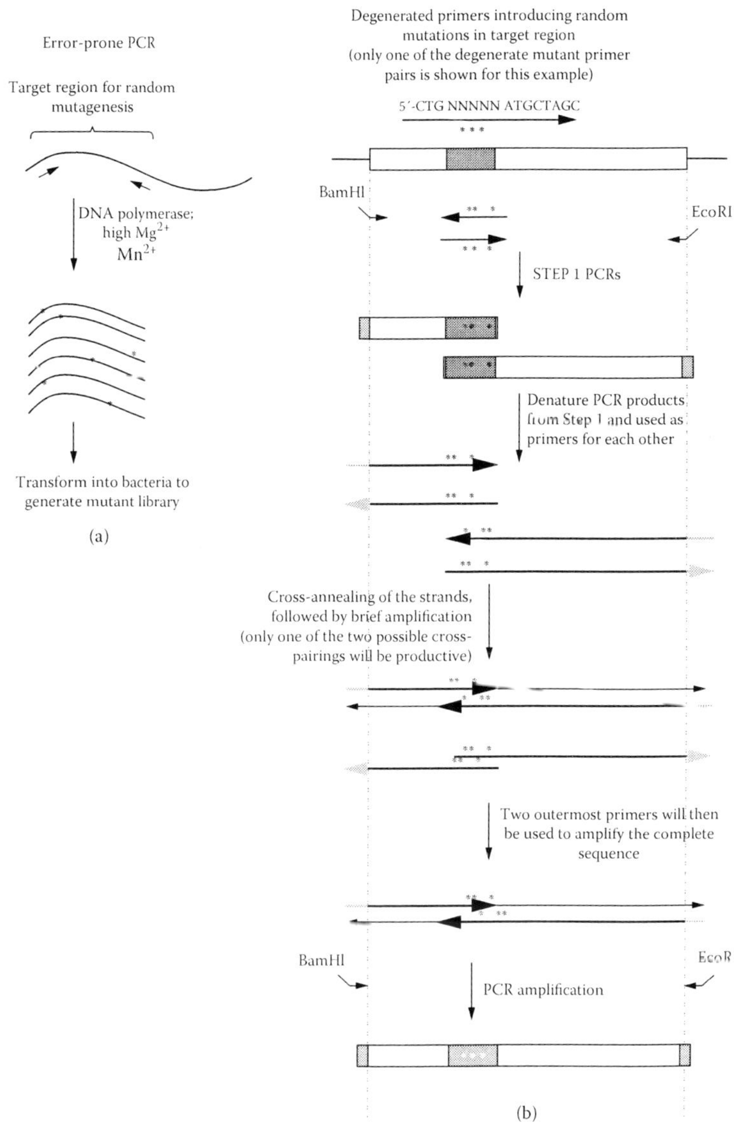

Figure 5.4 Two of the random mutagenesis strategies. (a) Error-prone PCR simply relies on the loss of specificity of certain DNA polymerases in the presence of high amounts of magnesium and manganese ions. (b) Site-directed random mutagenesis is essentially based on the same principle as site-directed mutagenesis, with the single difference of using degenerate primers across the target region. See the text for details.

generations have to be obtained for the "survival of the fittest." These molecules represent the engineered protein or the enzyme. The success of the directed evolution strategies therefore depends on how large the initial mutant library was, how many generations have passed, and which selection strategies were employed.

The strategy explained above is the traditional bioengineering approach that relies on existing genes, genetic networks, and protein architectures (Romero and Arnold 2009). The new field of synthetic biology is now offering previously unimaginable new combinations that do not exist in nature, by the designing and implementation of novel synthetic genetic circuits that will be discussed further (Dougherty and Arnold 2009; Koide 2009), see Section 10.2.

5.6 Problem Session

Answers are available for questions with an asterisk—see Appendix D.

Q1. Ins(1,4,5)*P*3 3-kinase B (IP3K-B) was shown in a study to colocalize to the cytoskeleton and be targeted to F-actin in the cell. Several mutants of IP3K-B were constructed for use in the experiment.

(a) Explain how these constructs were most likely generated.
(b) What is a *helix-breaking mutation*?
(c) How can you be sure that these deletion constructs are properly translated and folded?

Q2. Below is the coding sequence of the human H-Ras gene (introns are removed from the sequence):

atgacgg aatataagct ggtggtggtg ggcgccggcg gtgtgggcaa gagtgcgctg accatccagc tgatccagaa ccattttgtg gacgaatacg accccactat agag//

```
//gattcctac cggaagcagg tggtcattga tggggagacg tgcctgttgg acatcctgga
taccgccggc caggaggagt acagcgccat gcgggaccag tacatgcgca ccggggaggg
cttcctgtgt gtgtttgcca tcaacaacac caagtctttt gaggacatcc accagtacag//
  //gagcag atcaaacggg tgaaggactc ggatgacgtg cccatggtgc tggtggggaa
caagtgtgac ctggctgcac gcactgtgga atctcggcag gctcaggacc tcgcccgaag
ctacggcatc ccctacatcg agacctcggc caagacccgg cag//  //gagtggagg
atgccttcta cacgttggtg cgtgagatcc ggcagcacaa gctgcggaag ctgaaccctc
ctgatgagag tggccccggc tgcatgagct gcaagtgtgt gctctcctga
```

Below is its translation:

MTEYKLVVVG AGGVGKSAL TIQLIQNHFVDEYD PTIEDSYRKQV
VIDGETCLLD ILDTAGQEEYS AMRDQYMRTGEGF LCVFAINNTK
SFEDIHQYRE QIKRVKDSDDVP MVLVGNKCDL AARTVESRQAQD
LARSYGIPYIE TSAKTRQGVED AFYTLVREIRQH KLRKLNPPDE
SGPGCMSCKCVLS

Find out the domain structure of Ras, and design an experiment to generate Ras mutants defective in Sos binding but not on GTP binding.

***Q3.** Oncogenic RAS (the most upstream member of the MAPK pathway) is known to inactivate the BRCA1 DNA repair complex by dissociating BRCA1 from its binding sites on chromatin. You suspect that RAS can carry out this function by downregulation on the expression of BRIP1, a physiological partner of BRCA1 in the DNA repair pathway. How would you show that (a) Ras indeed downregulates a BRIP1 expression in breast tumor cells, and (b) that BRIP1 downregulation is necessary for the release of BRCA1 from chromatin?

Q4. You have identified the protein sequence of a novel cell cycle-related gene shown below:

MTEYKLVVVGA GGVGKSALTIQLIQNH FVDEYDPTIEDSY RKQVVIDGET 50
CLLDILDTAGQE EYSAMRDQYMRTGEG FLCVFAINNTKSFE DIHHYREQI 100
KRVKDSEDVPM VLVGNKCDLPSRTVDTK QAQDLARSYGIP FIETSAKTRQ 150
RVEDAFYTLVREIRQYRLKKISKEEKTPGCVKIKKCIIM 189

(a) Design an experimental procedure to obtain the *coding gene sequence* for this mature protein, and to clone it into a pGEX-2T bacterial expression vector.

(b) You have also hypothesized that phosphorylation can activate this protein. How would you design an experiment to identify which amino acid residues could be phosphorylated upon stimulation? Explain.

(c) Can you predict what the location of this protein would be in the cell? Explain the basis for this prediction. How would you experimentally show whether this prediction is correct? Describe the constructs required for this experiment and how you have constructed them.

Q5. You have identified a novel metabolic enzyme, NTR, which is expressed in aging human adipocytes, and cloned it into a pBluescript vector (shown in the figure). You have previously also obtained the promoter sequence for this gene and cloned it into a luciferase reporter vector (see Chapter 3, Section 3.5, Q6).

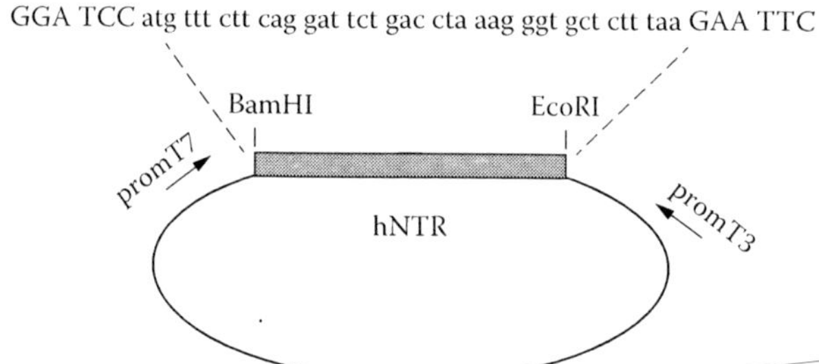

You now wish to analyze how the promoter of this gene is regulated in the young versus old mouse adipocyte cell, and for this purpose you want to identify the transcription factors regulating NTR expression, as well as to which sequence motifs these transcription factors bind on the promoter. How would you design your experiment?

Chapter 6

Protein–Protein Interactions

> The ideal engineer is a composite. He is not a scientist, he is not a mathematician, he is not a sociologist or a writer; but he may use the knowledge and techniques of any or all of these disciplines in solving engineering problems.
>
> **N.W. Dougherty (1955–)**

Introduction

One important feature associated with the function of a protein is its binding partners under different conditions. Some proteins have *obligatory* binding partners, which they cannot function without. Some other proteins make *transient* interactions depending on the situation. No matter what the context, these interactions are fundamentally important for the proper functioning of the cell.

There are many different methods to study these interactions, be it *in vitro* or *in vivo*. You can also choose your method of study depending on whether you have a hypothesis as to what these interaction partners are (i.e., have an "educated guess" and a potential target), or you have no *a priori* information and want to carry out a blind study.

As new techniques are constantly being developed, the ones that we will discuss here may be out of fashion, however, the results from these assays will nevertheless be valid.

6.1 GST Pull-Down-Based Interaction Assay

This assay was briefly mentioned in Chapter 4 (Section 4.3). In this assay, essentially the affinity between glutathione S-transferase enzyme and its substrate, glutathione (GSH) can be used to semipurify any protein X that is fused to GST by collecting it on glutathione beads. In this section, we will not only elute the protein from these beads; but we will also, use the proteins trapped on the beads to "fish out" any interaction partners, and analyze the presence or absence of the protein we suspect of interacting with it.

Just like creating the GFP fusion protein described in Section 4.8.4 in Chapter 4, care must be taken when cloning GST fusion proteins. Start and stop codons in the coding sequence as well as the reading frame must be controlled when designing a cloning strategy. Afterward, depending on whether the GST vector is for a bacterial or mammalian expression, the recombinant plasmid DNA must be transferred to the target cell, and the expression of the fusion protein confirmed. (In order to do this, you will have to use basic GST plasmid without any insert as an empty control plasmid, which will only express the GST enzyme, along with the recombinant construct. The correct protein size of the fusion protein should be estimated, and controlled by SDS-PAGE.)

Once the presence of the fusion protein in the lysates is confirmed, this lysate is then incubated with glutathione (GSH-coated) beads for some period, followed by pull down (hence,

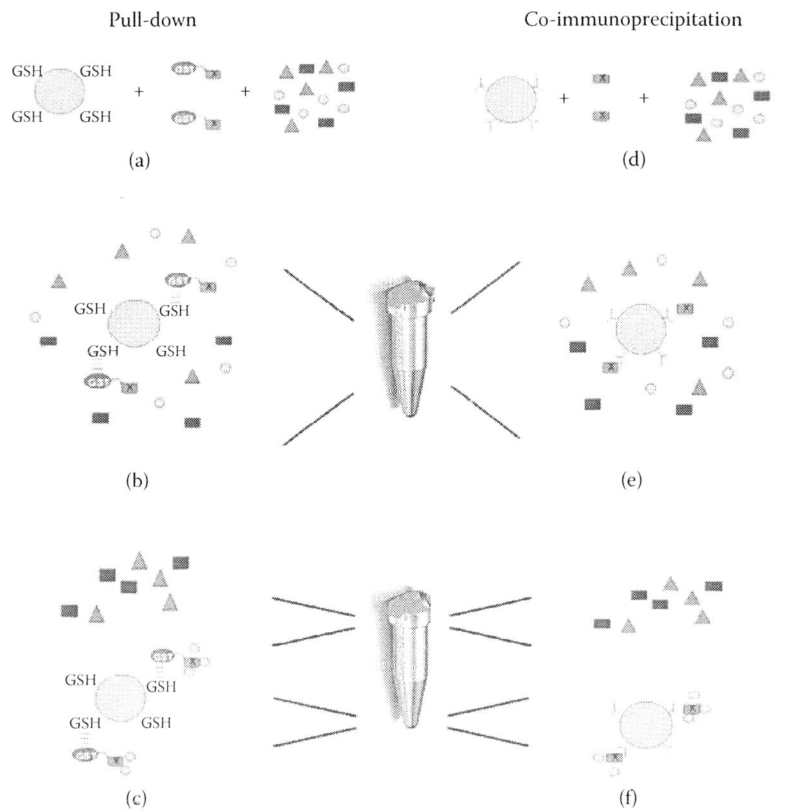

Figure 6.1 A schematic diagram comparing a GST pull-down assay (a,b,c) with a co-immunoprecipitation assay (d,e,f). Protein extracts containing either GST-fusion proteins (GST-X) or tagged or endogenous proteins (X) are mixed with either glutathione (GSH)-coated or antibody-coated beads in (a) and (d), respectively. These are incubated in a test tube for effective binding in (b) and (e). When the beads are pulled down or precipitated, both protein X (either GST-fusion or tagged or endogenous) and any interaction partner (shown here as the smaller gray circles) will also be pulled down (c) and (f).

the name). The GST-fusion protein will be precipitated with the beads due to an enzyme–substrate affinity, along with any other protein that may interact with the GST-fusion protein (Figure 6.1). The pulled-down proteins are then analyzed by Western blotting.

6.2 Co-Immunoprecipitation

The co-immunoprecipiation (Co-IP) assay is in principle very similar to the GST pull-down assay just described, with a slight difference; instead of a fusion with an enzyme, either the gene of interest is cloned into a tagged-vector (such as pCMV-HA described mostly in this book) or the endogenous gene product is used, and these proteins are precipitated using antibody-coated beads (Figure 6.1). The antigen–antibody affinity will be exploited for precipitating the protein of interest, along with any possible interaction partners.

6.3 The Yeast Two-Hybrid Assay

The yeast two-hybrid (Y2H) system, as well as the mammalian two-hybrid that follows the same principle, relies on the fact that transcription factors are modular, containing a DNA-binding domain and an activation domain (at a minimum) that are functionally separable units. The standard example in use in many Y2H vectors is the GAL4 protein in yeast and its cognate *gal4*-binding site on DNA (Figure 6.2a).

This assay can either be used to analyze the interaction between two known proteins (X and Y in Figure 6.2b), or to identify novel interaction partners of a protein X from a cDNA library. The proteins tested for interaction do not have to be transcription factors or even nuclear proteins at all; any intracellular soluble protein can be tested for interaction. However, the system is not suitable for analysis of membrane protein interactions (a different adaptation of the system is used for that purpose). Furthermore, the interaction is identified through the transcription of a reporter gene, most commonly *lacZ*, and a subsequent color reaction using substrate X-gal.

In order to use this assay, proteins X and Y have to be cloned into the appropriate vectors. Protein X is fused to the DNA-binding domain of GAL4, which will bind to the *gal4*

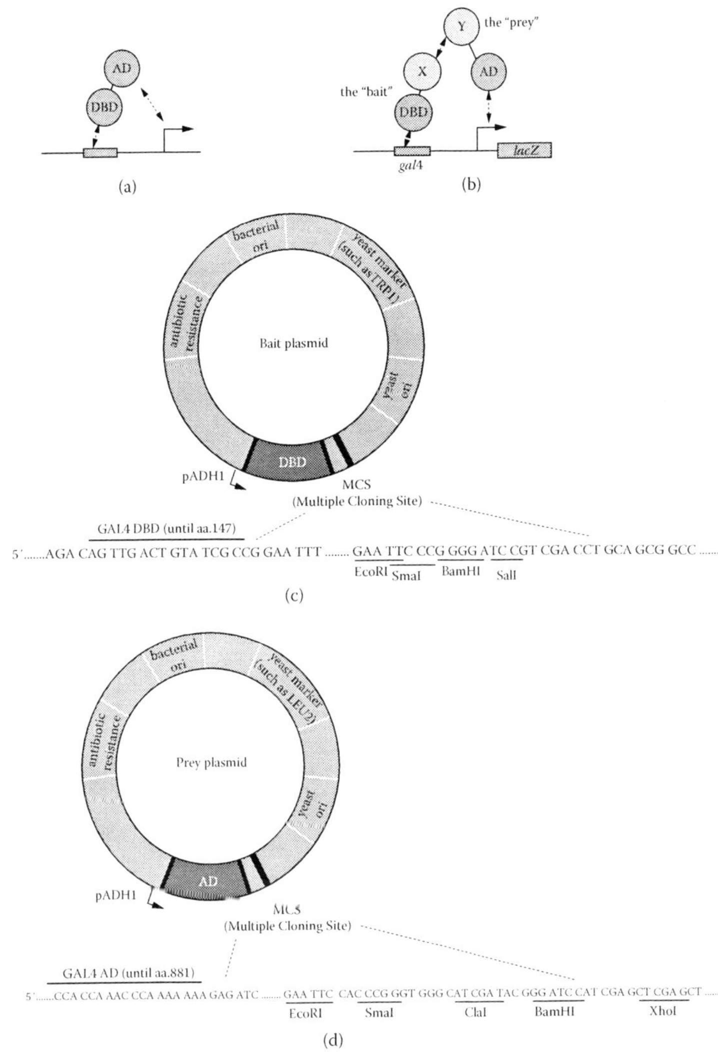

Figure 6.2 **A schematic diagram showing the basic principle behind the yeast two-hybrid (Y2H) system. (a) The modular structure of transcription factors, with a DNA-binding domain (DBD), and an activation domain (AD), and a cognate recognition motif on DNA (shown with a box). (b) The principle behind the Y2H system, where a protein X is fused to a DBD (typically that of GAL4), and a protein Y is fused to an AD (typically of GAL4), and the interaction between proteins X and Y is monitored through the expression of a reporter gene, such as *lacZ*. (c) The vector map for a typical bait vector. (d) The vector map for a typical prey vector.**

motif on the promoter that drives reporter expression, and hence it is called the **bait plasmid** (Figure 6.2c). Protein Y will then be fused to the activation domain of GAL4, which will be "caught" or "fished" by the GAL4DBD-X fusion protein, and hence it is called the **prey plasmid** (Figure 6.2d). The interaction is determined through expression of a reporter gene, usually stably integrated into the genome of a yeast strain.

Yeast is a unicellular eukaryote that has both haploid (asexual) and diploid (sexual) states, which is exploited in Y2H assays. In the haploid state, it can be found in either one of the two mating states: mating type a, or mating type α. The pheromones secreted from each mating type can attract these opposite-type haploids together (shmooing), which results in mating and the consequent diploid state (Figure 6.3).

This event is exploited in Y2H assays: each plasmid is separately transformed into a haploid yeast of different mating types (one of which usually contains a reporter already integrated into its genome). Subsequent mating will bring the different components of the system together into a diploid yeast, and reporter analysis is used to determine if the two proteins interact inside a living cell (Figure 6.4a,b).

The proteins analyzed (X and Y) can both be known and cloned accordingly, or alternatively a more general and *blind*

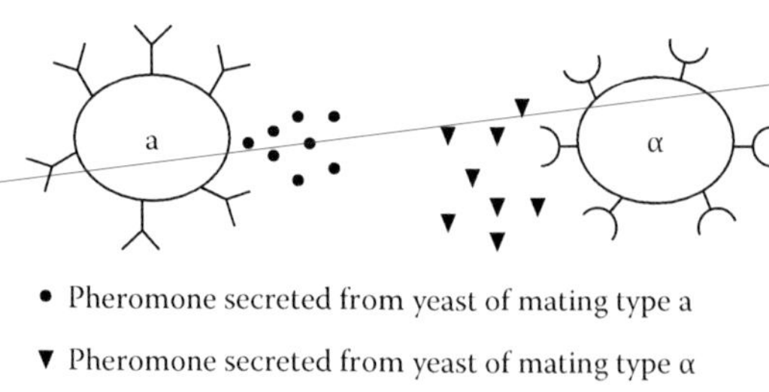

Figure 6.3 A summary of yeast mating. Haploids of opposite mating types a and α secrete pheromones that are detected by the other partner, thereby attracting each other. The end result will be mating and generation of an a/α diploid.

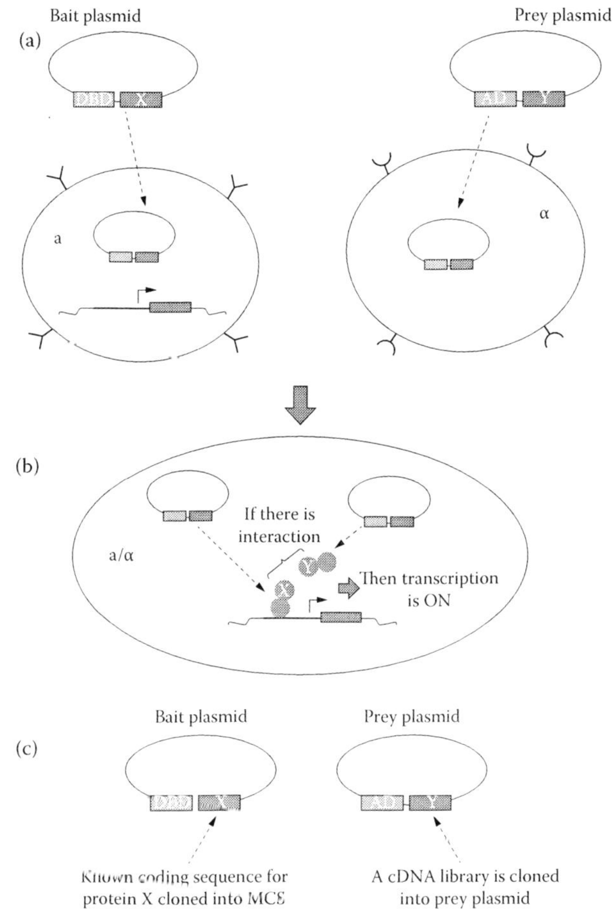

Figure 6.4 A summary of the yeast two-hybrid assay. Typically, haploids of opposite mating types are separately transformed with bait and prey plasmids (a), each encoding a DNA-binding domain (DBD) fused to protein X (bait) and an activation domain (AD) fused to protein Y (prey). One of the haploid yeast may already have a reporter plasmid integrated into its genome (called a *reporter yeast strain*). When the yeast of opposite mating types are mated (b), the resulting diploid will contain all three components of the yeast two-hybrid: bait plasmid, prey plasmid, and the reporter gene. If proteins X and Y interact, this interaction will bring the activation domain in close proximity of the promoter element driving reporter gene expression, and transcription will be switched on. (c) A cDNA library can be cloned into prey plasmid to carry out an interaction screening for all potential interaction partners of protein X.

study can be conducted to identify any interaction partner for protein X (i.e., protein X is known and cloned into the bait plasmid; however, the second component, Y, is unknown and will be fished out from a cDNA library cloned into prey plasmids) (Figure 6.4c).

One point that should be made is that antibiotics will not work as a selection pressure in yeast (most antibiotics are of fungal origin); therefore, another means of selection is needed. Nutritional supplements and metabolic enzymes come in handy in that respect. Wild-type yeast can normally grow on a minimal medium that contains basic essential ingredients and the *essential amino acids*, which yeast use to synthesize all other amino acids themselves. When yeast is mutant for one of the necessary synthetic enzymes, however, they can no longer convert these essential amino acids to others necessary for growth, and need nutritional supplements (Figure 6.5). This event is exploited in yeast genetics for selection.

Several spin-offs, or variants, of this Y2H analysis have been devised, such as yeast one-hybrid, yeast three-hybrid, mammalian two-hybrid, and so on, each with a different purpose and application.

Yeast one-hybrid (Y1H) was designed to monitor protein–DNA interactions, and most commonly to fish out any proteins that bind to a particular DNA motif (Figure 6.6a). Once again, a cDNA library has to be cloned into the prey plasmid and transformed into yeast. Expression of the reporter gene is driven by a minimal promoter activated by an upstream activating motif. The purpose of this Y1H assay is most commonly identification of all possible proteins that bind to this DNA sequence. The protein does not necessarily have to be an activator (it can be an inhibitor that binds to and masks this motif—one may try to identify proteins binding to an origin of replication, or proteins binding to insulator motifs, etc.); therefore, the library is again fused to an activation domain as in a typical prey plasmid.

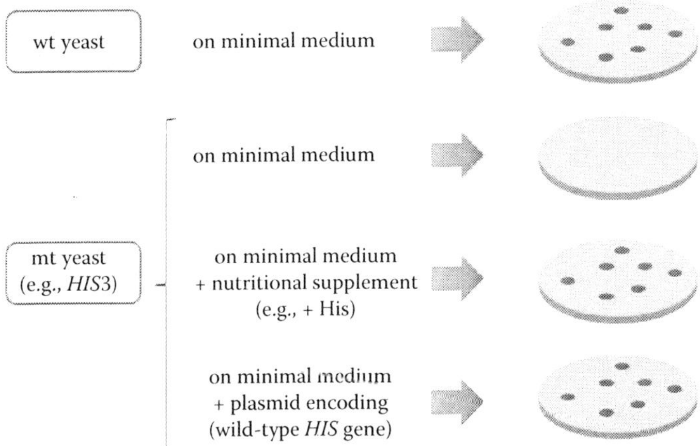

Figure 6.5 Yeast selection. Wild-type yeast contain metabolic enzymes necessary to synthesize all the required macromolecules from essential amino acids and other material, and can therefore survive on a minimal medium that only contains essential nutrients. A yeast mutant in one of these enzymes, such as the *HIS3* gene, however, cannot survive on a minimal medium and instead requires the missing amino acid as a nutritional supplement, or needs to be transformed with the plasmid containing the wild-type version of the gene, restoring normal metabolism.

A yeast three-hybrid assay (Y3H) is usually used when a three-way interaction is studied. For example, blockers of a receptor–ligand or enzyme–substrate interaction may be studied; the interaction of two RNA-binding proteins via an RNA molecule may be investigated; chemical modifiers of a protein–protein interaction may be studied, and so on. When large chemical or other molecular libraries are investigated for a blocking interaction, a simple reporter assay is not sufficient; high-throughput screening assays commonly, although not exclusively, exploit the use of death genes (i.e., cytotoxic genes) as reporters (Figure 6.6b). The mammalian two-hybrid follows a very similar strategy to that in the Y2H, so it is not discussed further.

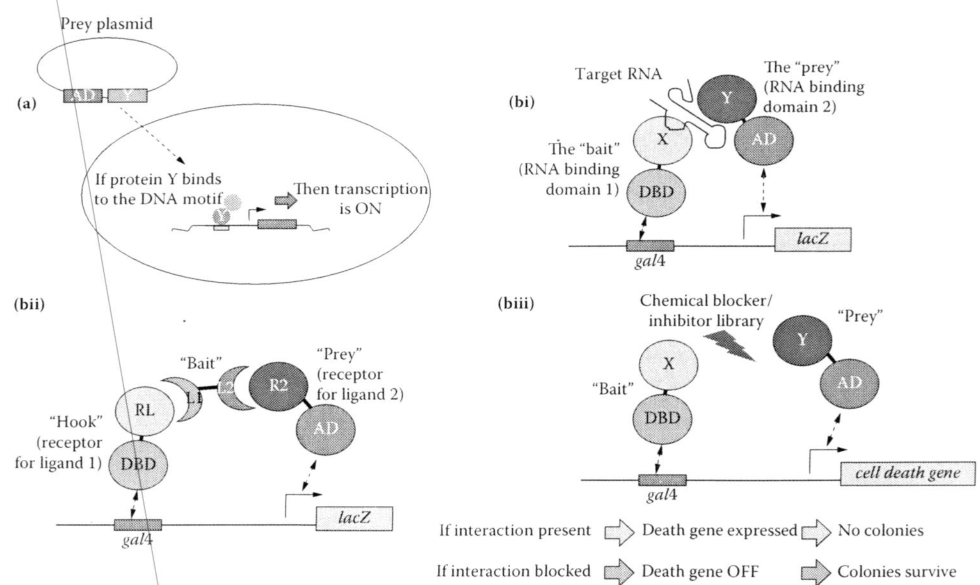

Figure 6.6 Yeast one-hybrid (a) and three-hybrid (b) systems. (a) In Y1H, the prey library is screened for any positive interaction with the target DNA-binding motif. (b) Yeast three-hybrid can be used for a number of different reasons, such as the (i) identification of three-way interactions between an RNA molecule and two RNA-interacting domains, (ii) analysis of two different receptors (R1, R2) with either the same (not shown) or different (L1, L2) ligands, or (iii) screening of a chemical library for potential blockers of an interaction between two proteins X and Y. For the latter case, high-throughput screening is made easier by the use of a cytotoxic gene (cell death gene) as a reporter.

6.4 Fluorescence Resonance Energy Transfer

In vitro interaction assays are highly artificial methods for analysis of protein–protein interactions; the yeast or mammalian two-hybrid system investigates interactions *in vivo*; however, the assay relies on end-point analysis and does not give much information about the dynamic nature of the interactions in real time. Therefore, new methods have been developed for such dynamic studies.

The most commonly used method for such a dynamic study is fluorescence resonance energy transfer (FRET). This method relies on the transfer of resonance energy when two fluorescent molecules are close enough to interact (several angstroms). However, one drawback is that both proteins under investigation have to be fused to different fluorescent proteins for such an energy transfer to occur, which in some cases may interfere with either the localization or function of the proteins.

The technique exploits the absorbance (excitation) and emission spectra of fluorescent molecules, as shown in Figure 6.7 (right panels). We will not go into the details of the physics of light here, however, it should be noted that the wavelength of light and its energy are inversely correlated; that is, the lower the wavelength, the higher the energy and vice versa. Thus, fluorescent proteins that are excited by a low wavelength light (which has to be within the excitation spectrum) (Figure 6.7a) will emit fluorescence that is of a higher wavelength (within the emission spectrum) (Figure 6.7a). The fluorescent microscopes (epifluorescent, confocal, or others) usually use the peak values of these spectra for excitation and emission, using appropriate filters.

Proteins whose interactions are under investigation are commonly cloned into expression plasmids that will create fusions to various fluorescent proteins (Figure 6.7a, left panels). The fusion protein expressed from this plasmid will be a fusion of the relevant fluorescent protein and the protein under

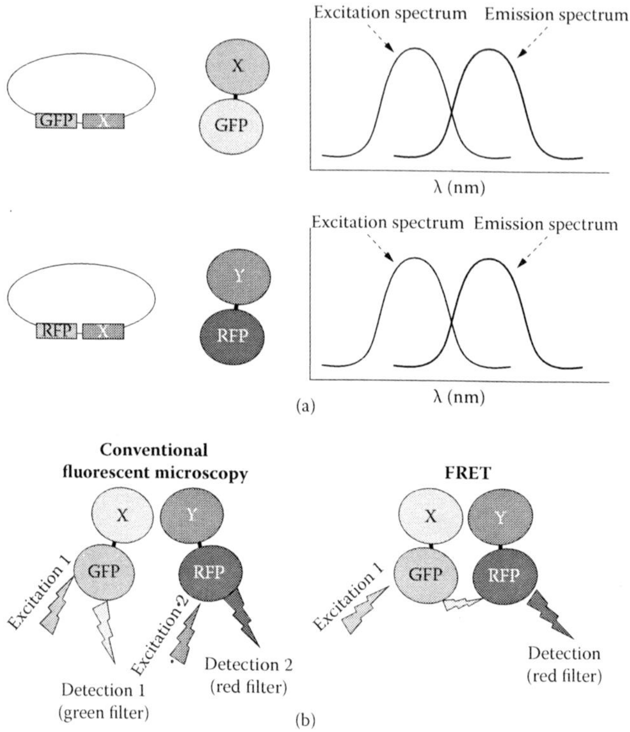

Fluorescent Protein	Excitation (nm)	Emission (nm)
mPlum	590	649
mCherry	588	610
DsRed	566	586
mOrange	548	586
EYFP	514	527
EGFP	488	527
mCFP	433	475
T-Sapphire	399	511

(c)

Figure 6.7 An overview of FRET analysis. (a) Cloning of both proteins X and Y to be investigated for interaction into appropriate fluorescent expression vectors, pEGFP and pDsRed, respectively (as an example, left panels), the resulting fusion proteins (middle panels), and the respective excitation and emission spectra (right panels). (b) Comparison of excitation and detection of fluorescence in conventional microscopy (left panel) versus FRET (right panel). (c) A sample table summarizing excitation and emission peaks (filters in fluorescent microscopes) of some common fluorescent proteins.

study (Figure 6.7a, middle panels). In conventional fluorescent microscopy techniques, be it epifluorescent or confocal fluorescent microscopy, the two fluorescent proteins that are co-expressed in the same cell are excited separately, using different excitation wavelengths, and the resulting fluorescence emitted from each protein is detected separately (Figure 6.7b, left panel). In FRET, however, only one of the fluorescent proteins is excited; the fluorescence emitted from this fluorescent protein will be transferred to the neighboring fluorescent protein (if the fusion partners are interacting, bringing the fluorescent proteins close enough for this transfer to occur); and the second molecule that is thus excited from resonance energy transfer will in turn emit fluorescence, which will be detected by the researcher (Figure 6.7b, right panel). Since excitation and emission of fluorescent proteins can be carried out in live cells, dynamic protein–protein interactions can be studied in real time.

The crucial point here is that the emission of the first fluorescent protein should have enough energy to excite the second one; that is, the emission spectrum of the first fluorescent protein must overlap with the excitation spectrum of the second one. Therefore, the design of the FRET experiment and the relevant clonings are the most important step in such a study. The standard fluorescent protein has been the enhanced GFP variant, EGFP, as discussed in Chapter 4, Section 4.8.4. Since the engineering of the first generation EGFP vectors, many variants of this protein have been developed, including mPlum, mCherry, ECFP, EYFP, EBFP, and so on; the excitation and emission peaks of which are exemplified in Figure 6.7c.

6.5 Problem Session

Answers are available for the questions with an asterisk—see Appendix D.

***Q1.** If you wanted to study the dynamic interaction between the two growth factor receptors JGF1 and JGF2, how would you design the experiment? Explain.

JGF1 5′-ATG CAG CCA TTT TGC CGG AGG TAG-3′
JGF2 5′- ATG CCC CAC TTA TGG CGG AGA TGA-3′

Q2. Following the same experiment in Chapter 5, Q5, how would you design an experiment to study if hNTR interacts with its coenzyme QXR in both the young and old adipocyte in the same manner? (Assume that you already have QXR cloned in a pGEX vector, but as of yet there is no commercial antibody for hNTR. How would you solve this problem?)

Q3. You have previously demonstrated in your lab that metaphase chromosome protein 1 (MCP1) is involved in the early events of DNA replication. You now want to show that MCP1 associates with proteins that are required for the establishment of the pre-replication complex such as ORC2, ORC4, and MCM2. The sequence of the MCP1 protein is as follows:

N-Met-Lys-Arg-Ala-Leu-Lys-His-Val-Arg-Arg-Cys-C

(a) There are no plasmids available yet for ORC2, ORC4, or MCM2, but commercial antibodies for all three can be purchased. How would you design this experiment?
(b) You then wish to carry out immunofluorescence to study the colocalization of MCP1 with some of those proteins. How would this assay be designed?
(c) Moreover, you also need to show that these proteins all form a complex on the replication origins (5′-ATGCTAATCATCTACCTAA) in human cells. How would you design such an assay?

Q4. Transfer RNA (tRNA) is an essential component of the cell's translation apparatus. These RNA strands contain the anticodon for a given amino acid, and when *charged* with that amino acid they are termed *aminoacyl-tRNA*. Aminoacylation, which occurs exclusively at one of the 3′-terminal hydroxyl groups of tRNA, is catalyzed by a family of enzymes called *aminoacyl-tRNA synthetases* (ARSs). You are instructed to select a novel ARS-like protein (you will term this as a *flexizyme*) from a random pool that will recognize the target 5′-leader region of tRNA (5′-UAGGCGAUCGAGCCCGAUCGA-) as a substrate. Devise a strategy.

Q5. You are very much interested in the regulation of a novel *UD1* promoter by transcription factor BE2. You know that a BE2 expression is induced in response to high Na^+ in the medium. How would you design an experiment to show that BE2 does indeed bind to the *UD1* promoter *in vivo*? Explain your strategy. (You already have the promoter cloned in pBS plasmid as follows, and the BE2 you use in the experiments will be the *endogenous protein*.) The figure shows a schematic diagram of the *UD1* promoter construct cloned in pBluescript plasmid (pT/only shows the orientation). The underlined text indicates the putative BE2 binding site.

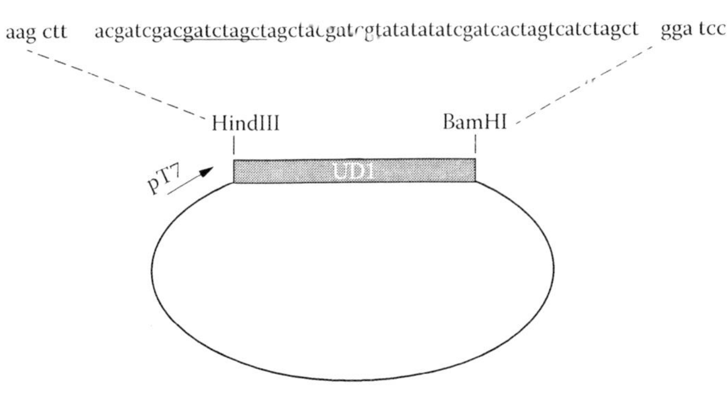

Q6. It is known that the Lys-Ala-Cys-Gln-Val residues (shown in boxed text in the figure) on BE2 are involved in binding to DNA motifs. Can you determine which one of these amino acids in particular is involved in regulating the *UD1* promoter elements? The figure shows a schematic diagram of the BE2 coding sequence cloned in a pCMV-HA plasmid. The underlined sequence shows the BE2 coding region, whereas the boxed sequence indicates the putative binding domain. (You have already cloned BE2 to the pCMV vector as shown below.)

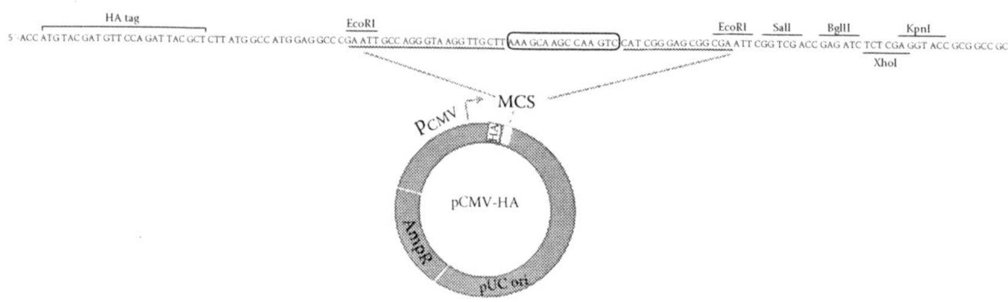

Q7. The *UD1* promoter drives the synthesis of an enzyme called **silicatein** that polymerizes silica nanospheres used in bioengineering and tissue engineering applications; however, its production and purification by classical methods have so far been difficult. Researchers propose that adding a glutamate-tag (multiple glutamates, at least six amino acids) could be more productive than an HA-tag in immobilization on hydroxyapatite columns.

(a) How would you engineer a pCMV vector series that contains multiple Glu-tags instead of the HA-tag for the purpose of cloning the *UD1* coding sequence?
(b) How would you clone the coding sequence of this enzyme into your engineered pCMV-Glu vector, if only the enzyme sequence is known as shown below:

N-Met-Ala-Gly-Gly-Leu-His-Ile-Cys-Ser-Pro-Met-Val-C

(You also know the *UD1* promoter; see Q5.)

Chapter 7

Cell Culture

> Some cultural phenomena bear a striking resemblance to the cells of cell biology, actively preserving themselves in their social environments, finding the nutrients they need and fending off the causes of their dissolution.
>
> **Daniel Dennett III (1942–)**

Introduction

Early cell culture studies mainly used cells isolated from a frog, where incubation was not required since frogs are warm-blooded animals. However, as investigations became faster and more complex, a model system which more closely resembled human systems was required. The first choice was rodents, which produced continuous cell lines and the transgenics were easier to achieve. Human cells, however, were more difficult to culture. This difficulty was overcome initially, with many thanks to Henrietta Lacks, who was descended from the slaves and lived on the plantation of her ancestors. Henrietta had a very aggressive tumor all over her body. The cells isolated from her cervical tumor samples propagated readily in culture systems, called *HeLa cells*, and they are still being used today, some 60 odd years later.

There are many different types of cultures, so let us first begin with some terminology. **Tissue culture** refers to the *in vitro* cultivation or *culturing* of an entire tissue, such as epidermal tissue of the skin. Obviously, due to intrinsic features of the cells such as their dividing capacities, some tissues are easier to grow *in vitro* than others. **Cell culture** refers to the *in vitro* culturing of dispersed cells derived from either primary tissue (in which case it is called **primary cell culture**), from a cell line, or from a cell strain (we will discuss the difference of these latter two terms later). **Organ culture** refers to the three-dimensional culturing of *undisintegrated* tissue (thereby preserving some of the features of the organ although it is detached from the body). **Organotypic culture**, on the other hand, refers to the recombination or coculturing of cells from different lineages *in vitro*, so as to recreate some of the features of the organ. And, of course, all of the above terminologies may refer to animal cells, plant cells, or insect cells.

A primary cell culture is derived directly from tissue and therefore best represents the tissue of interest, however, these cells have a limited life span *ex vivo* (i.e., outside the organism). A primary cell culture is also difficult to obtain, simply because each tissue is composed of multiple cell types and thus is not homogenous—the cell type of interest has to be seperated from others in the tissue—and furthermore, the tissue is commonly swamped by extracellular matrix proteins onto which cells are tightly attached, in addition to cell-to-cell interactions. Therefore, the matrix proteins first have to be degraded, then the cells have to be disaggregated by mechanical or enzymatic means (with the exception of blood cells which are already dissociated), viable cells of interest separated, and then cultured with the help of essential growth factors and supplements, all while avoiding any contamination. Primary cells can be grown for multiple generations in the laboratory, giving rise to cell strains, however, these have to be validated in many aspects before they can be classified as cell lines.

Cell lines are simply transformed or otherwise immortalized cells that can be maintained *in vitro* for many generations without much change in physiology. These cells can either be primary cells that have been engineered to express viral oncogenes and thus exhibit unlimited growth capacity, or derivatives of tumor cells that are immortalized.

Therefore, depending on what type of culture one wishes to use in one's studies, there are some important issues to be considered, such as (a) the source of cells; (b) the dispersion methods (not every tissue can be dispersed in the same way, due to differences in matrix material); (c) the defined media (i.e., antibiotics, nutritional supplements, growth factors, hormones, etc.); (d) temperature; (e) pH; (f) incubators; and (g) the method of propagation of cells over passages.

Cells can be grown as either adherent or suspension cultures, depending on the source or origin: for example, fibroblast or epidermal cells require tight attachment or adherence to an extracellular matrix, and they have to be grown as adherent cells. On the other hand, leukocytes (white blood cells) are cells that normally circulate in the bloodstream without any significant attachment, and are highly mobile, therefore they can be grown as suspension cells. Adherent cells are normally maintained as monolayer cultures (i.e., grown as a single layer, and then growth will be stopped by contact inhibition, or cells will be passaged beforehand), however, most cell lines are immortalized by oncogenes and are therefore incapable of contact inhibition. This means they can be overgrown once they reach confluency; therefore cells have to be passaged (or diluted) on a regular basis before they reach this state. Suspension cells should also be passaged, simply because the nutrients in the medium will not suffice in a rather crowded population of cells and will eventually have to be replenished.

Thus, the advantage of using tissue culture is control of the physiochemical environment, since semidefined media are used, where nutrient concentrations can be determined, but the content and amount of growth factors and other macromolecules within sera such as fetal calf serum or horse serum cannot be determined and there are changes in different batches. Furthermore, defined incubation conditions (such as temperature, CO_2 concentration, etc.) can be employed. The cell lines or even cultures from primary tissues are relatively homogenous and after a few passages become more uniform (largely due to selective pressure or culture conditions); and cells can be used almost indefinitely, especially true for cell lines, if cells are stored in liquid nitrogen. When experiments are carried out, the cells in culture are usually directly exposed to the chemicals or reagents tested—this can also be considered a disadvantage, since in the organism the chemicals or reagents are metabolized and distributed to tissues through blood.

The disadvantages, however, are that special expertise and highly aseptic conditions are required, which are not readily available at all times. Animal cells grow slowly as it is, and the high risk of contamination by bacteria, yeast, mold, or fungi complicates matters. The amount of cells that can be grown under these circumstances is very limited; moreover, cells can change their characteristics, such as losing their differentiated status, becoming more aggressive, or showing signs of senescence, when maintained for too long in culture.

7.1 Genetic Manipulation of Cells

The method for genetic manipulation of cells is largely dependent on the type of cell in question; however, one can subdivide these methods into five major categories:

1. *Electrical*, such as in electroporation
2. *Mechanical*, such as in microinjection or a gene gun
3. *Chemical*, such as with liposomes or calcium phosphate
4. *Viral*, such as with baculoviral, retroviral, or lentiviral vectors
5. *Laser*, such as in phototransfections

The expression plasmids that were discussed in Chapter 2 onward can be delivered to the target cells using any one of these methods, which will now be described in detail.

Despite the method employed for the transfer of DNA into the target cell, not every cell will receive this foreign DNA. Hence, we have to define the term **transfection efficiency**, which is essentially the percentage of cells that are successfully transformed among the total cell population. Most commonly, a reporter such as GFP is used to optimize the transfection by the calculating transfection efficiency obtained under different conditions:

$$\text{transfection efficiency} = 100 \times \frac{\text{transfected cell number}}{\text{total cell number}}$$

Each cell type will have a different transfection efficiency under each method, influenced also by the DNA amount or reagent used among many other parameters; therefore, before each experiment the transfection conditions have to be optimized for that cell type. The confluency of the cell is also important for transfection efficiency: usually cells at 40 to 70% confluency yield better transfection efficiencies when compared to cell populations that are too dense or too sparse.

7.1.1 Electrical Methods

Electroporation of animal cells essentially works with the same principle as electroporation of bacteria that was introduced in Chapter 2 (see Figure 2.26c), where the cells suspended in an appropriate solution are subjected to an electrical field, which causes transient openings in the cell membrane through which DNA that is present in the solution can penetrate the cell. This is, however, a risky operation, and there must be a delicate balance between the intensity of the electrical field and the duration so as not to disrupt the integrity of the cell membrane.

7.1.2 Mechanical Methods

This method of DNA transfer involves direct interference with the membrane integrity through mechanical means, such as microscale needle, that is, microinjection, and gene gun. These particularly come in handy when there are very few cells to be transfected, or when the cell is structurally difficult to penetrate, such as a plant cell due to its cell wall.

Microinjection is a technique for transfer of DNA to cells using a glass micropipette that can directly penetrate cell membranes, cell walls, or even nuclear envelopes as required. This procedure is carried out with a special instrument called the *micromanipulator* under an optical microscope, where the target cell is immobilized with the help of a blunt pipette as the microinjection needle or the glass micropipette delivers the DNA. This technique can even be used to transfer DNA into the male pronucleus of a fertilized egg for creating transgenic animals.

A gene gun, also called a *particle gun* or *biolistic system*, was originally designed for the effective transfer of DNA into plant cells, whose cell walls are hard to penetrate by chemical techniques. Essentially, a heavy metal such as gold or tungsten that is coated with the DNA to be transferred is accelerated under helium or a similar propellant, and as the rather fast metal penetrates the cell wall and the cell membrane, it also delivers the DNA into the cell. If membrane integrity is severely compromised, however, the viability of the cells after the operation (hence the transformation efficiency) may be low.

7.1.3 Chemical Methods

Chemical methods involve the precipitation of DNA using different chemicals and uptake of this mix by the cell. The commonly used method involves calcium phosphate cotransfection.

Calcium phosphate-mediated transfection is a rather inexpensive method. Basically, the plasmid DNA is allowed to form a precipitate in the presence of calcium ions in the solution (calcium chloride) followed by a phosphate buffer addition, which is thought to mediate uptake of this precipitated DNA by the cell through endocytosis or phagocytosis. The disadvantage of this method is the relatively high degree of variability among parallel experiments.

Liposome-mediated transfection can also be classified as a chemical method, since it involves the encapsulation of DNA by lipid molecules with hydrophilic modifications (for easy dissolution and dispersion); the lipid component makes it easy to integrate with the cell membrane. The liposome is essentially a lipid bilayer, and as such can entrap DNA molecules within them (or even drugs, depending on the application); if these liposomes are formed by cationic lipids, the positively charged head groups will interact better with the negatively charged DNA molecules. Since the lipid bilayer of the liposome is compatible with the phospholipid bilayer of the cell membrane, allowing for either the fusion of the liposome with the membrane or endocytosis, liposomes generally offer a high transfection efficiency for animal cells. The DNA that it thus transferred to the cell first appears in the endosomes and later in the cell nucleus. However, the presence of a cell wall on plants or bacteria makes it difficult for liposomes to deliver DNA to these cell types.

7.1.4 Viral Methods

Particularly for the purposes of stable transfections and integration into the genome and for *in vivo* delivery, viral vectors are commonly used. A number of different viruses have been exploited for their ability to transfer DNA into target cells, however, one thing remains common: viruses have host specificity, that is to say every virus has a particular cell type that it can normally infect. Therefore, a different type of viral vector may be required for each cell type as appropriate. Viruses are infectious agents, however, when exploited to transfer our DNA of choice, the idea is that the DNA stays in the target cell—that is, the viral vectors, once they deliver the DNA to the target cell, should not form new viral particles and lyse the cell. Therefore, again common to all viral vectors, infectious genes are usually removed from the viral genome, and only those that are required for a single round of infection are maintained. This is also the disadvantage (or rather, difficulty) of engineering and working with viral vectors.

The common property of all viral vectors is that they have been engineered so that they are no longer infectious, that is, once they deliver the cargo DNA to the target cell they cannot produce any more viral particles. However, this biosafety measure also brings a logistic problem along: how to package the cargo DNA into viral particles in the first place. To this end, the commonly employed method is to use a "packaging cell line," which contains some of the crucial viral genes that are stably integrated into its genome. In more recent versions of viral vectors, some of these genes are further divided among packaging cell lines and *helper vectors* as an additional safety measure (such that the viral particles can only be packaged if three or more components are present simultaneously) (Figure 7.1).

Once the packaging cells produce viral particles, one needs to quantify the amount of virus obtained so that target cells can be effectively infected by the appropriate viral titer, which is a functional measure of viral infectivity. The most common method used to calculate the viral titer is the plaque assay, whereby plaque-forming units (PFUs) can be calculated. Simply, cells are grown to confluency in multiwell plates, and serial dilutions of the viral sample are applied onto the cells. The virus that infects a cell will lyse and spread to the neighboring cells, thereby creating a *plaque* (Figure 7.2). The number of plaque are counted in different wells, and a plaque-forming

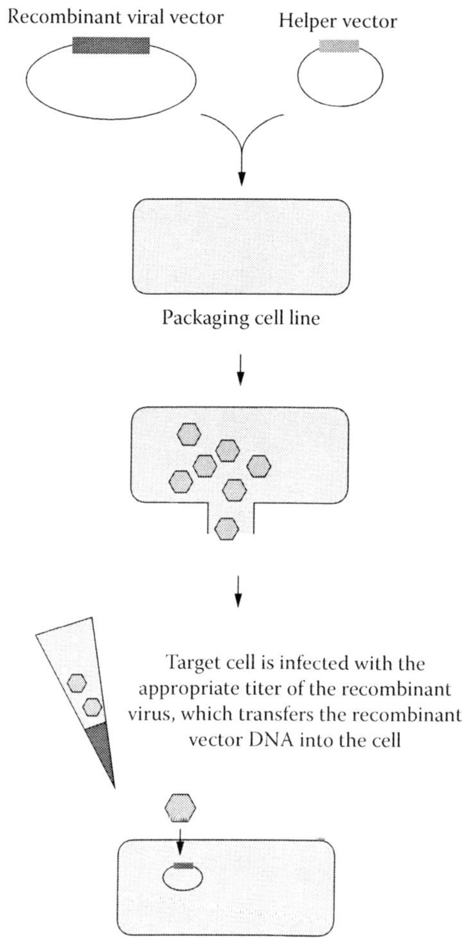

Figure 7.1 A simplified scheme of how viral vectors are typically used. The viral vector that is constructed to carry the insert DNA is co-transfected to a packaging cell line along with a helper vector. The recombinant DNA gets packaged into viral particles; after calculation of a viral titer (see Figure 7.2), target cells are infected with the recombinant virus, allowing for the transfer of recombinant DNA.

unit per sample volume is calculated (PFU/ml). This information is then used to calculate how much of the virus should be used to infect the target cell.

Having said this, viral expression systems that have been recently developed may not necessarily require a plaque assay;

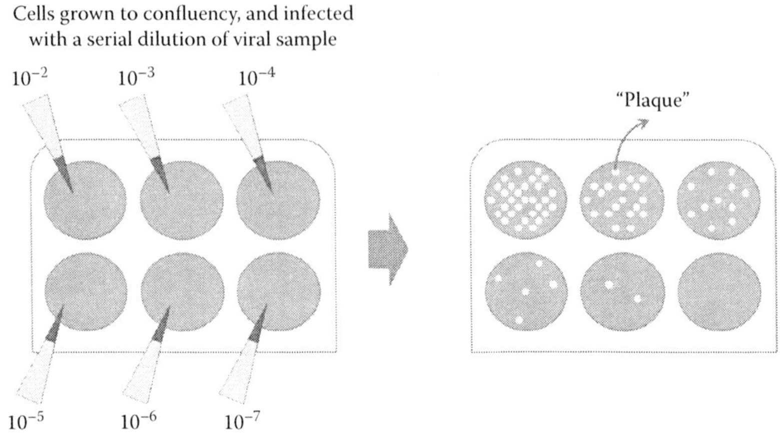

Figure 7.2 A simplified summary of a viral plaque assay. Cells are grown to confluency and infected with serial dilutions of the viral sample (e.g., 1 in 100, 1 in 1000, etc., dilutions, as indicated in each pipette tip above). After an incubation period (specific for each virus), the infected cells get lysed and the released viruses spread to neighboring cells, a *plaque* appears around each original infection site. The numbers of plaque can be counted, and used in calculation of the plaque-forming unit (PFU) per ml of sample.

however, since different companies produce different plaque-free systems, they will not be discussed here.

Adenoviruses have a double-stranded linear DNA as their genome, and no viral envelope. They can infect a number of different vertebrates including humans, thus they present a valuable resource as gene therapy vectors for clinical use. Adenovirus-based vectors can therefore infect human cells as well as many other mammalian model cell lines (mouse, rat, etc.), and can accomodate rather large DNA inserts, and transfer them into both dividing and nondividing cells (e.g., for use in gene therapy for Parkinson's disease). However, the packaging capacity is still lower than other viral vectors such as retro- or lentiviral vectors. Adenoviral vectors typically contain a replication-deficient partial genomic sequence, and may require a helper vector that contains part of the genome, or that undergoes homologous recombination with the recombinant vector, so as to achieve production of adenoviral particles (Figure 7.3a).

Baculoviruses are rod-like viruses with a circular double-stranded DNA genome, and they typically infect invertebrates, particularly insects such as moths or mosquitoes. Baculoviral vectors are therefore used to transfer DNA into insect cells such as Sf9 cells, mostly with the intention of expressing proteins in large quantities for further analysis. Baculoviruses are preferred to other viruses if merely a large-scale expression of proteins is required, simply because of their nonpathogenicity: meaning, baculoviruses cannot infect mammals or plants. The coding sequence for the protein is usually cloned under the polyhedrin promoter, which normally transcribes high levels of the baculoviral polyhedrin protein toward the end of the infection, thus a large-scale and high-level expression of proteins can be achieved. There are many different systems for baculoviral expression, such as the *bac*-to-*bac* system (Invitrogen) and the *flash*-bac system (Oxford Expression Technologies), but they all essentially rely on the strong expression of the gene of interest (GOI) from the viral *polyhedrin* promoter (see Figure 7.3b and Figure 4.4 in Chapter 4).

Retroviruses are RNA viruses and use the reverse transcriptase enzyme (which was discussed in Chapter 3) to generate DNA from its RNA genome for integration into the host chromosomes. Thereafter, the virus transcribes RNA as its genome, as well as expresses proteins necessary for packaging. These proteins include polymerase (pol) that codes for the reverse transcriptase, group-specific antigen (gag), protease, integrase, and envelope (env) proteins. Retroviral vectors are therefore based on this genome, and contain the long terminal repeats (LTR), psi (Ψ) required for packaging, and the constitutive CMV or SV40 promoters driving the expression of the gene of interest. There are many variations of these vectors, using either constitutive or inducible promoters (described later), different packaging or helper vectors, and so on, but a generic vector set is shown in Figure 7.3c.

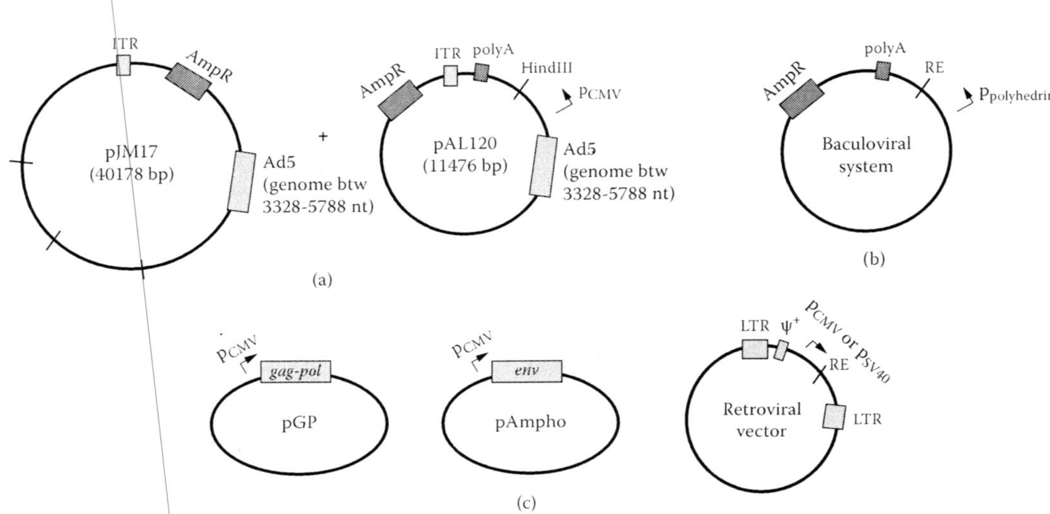

Figure 7.3 Schematic representations of some common viral vectors. (a) Adenoviral vector pAL120 and helper vector pJM17, based on Adenovirus type 5. ITRs are inverted terminal repeats; pJM17 vector contains the replication-defective portion of the Ad5 genome (note that there are various HindIII sites within the plasmid, shown as short lines). The pAL120 vector is used to clone the gene of interest into the HindIII site, under the pCMV promoter. Upon coinfection, homologous replication will transfer the gene of interest to the Ad5 genome in pJM17, in place of the E1 gene, and allow for adenoviral particle production. (b) Baculoviral vectors are used typically to express genes under the polyhedron promoter of baculoviruses. (c) Retroviral vectors (far right) are usually used to clone genes under the constitutive CMV or SV40 promoters, and include the psi-packaging elements (Ψ+); they, however, require the presence of other helper vectors that express gag, pol, and env proteins (see the text for details). LTR: long terminal repeats.

The disadvantage of retroviral vectors is that they are replication competent, that is, they only infect cells that are actively dividing, and may also activate any latent disease in the target cells or endogenous retroviruses. However, when quiescent primary cells are used (such as a myocyte) then recombinant retroviral particles that have thus been obtained will not be useful. For such cell types, lentiviral vectors are typically the choice.

Lentiviral vectors are based on lentiviruses, a subclass of retroviruses, but they are particularly exciting for molecular research since they can infect nondividing quiescent cells (in G_0, such as neurons or muscle), that is, the *preintegration complex* or the *virus shell* can penetrate the nuclear envelope, and thus allow for stable integration of the insert DNA into the genome.

7.1.5 Laser Methods

Phototransfection or optical transfection is based on the principle that when a laser beam is focused on a given spot on the cell membrane, a transient pore is generated through which nucleic acids, DNA or RNA, can be transferred to the cell. This is particularly useful when a limited number of primary cells will be analyzed (for example, single cell analysis of cortical neurons), however, it is not as applicable to a large number of cells. Nonetheless, it presents the advantage that mRNAs can be transferred to particular cellular locations for local translation, which cannot be achieved using most of the above-mentioned methods.

7.2 Reporter Genes

We already reviewed expression plasmids in Chapter 2; therefore, we will not discuss the details of those vectors again here. We will only remind the readers that the plasmids used for transfection and analysis in (animal) cells broadly fall into four different catergories based on the promoters used:

1. Plasmids containing minimal promoters, such as the HSV *tk* (herpes simplex virus *thymidine kinase*) promoter, are essentially used to study enhancer elements or activating regulatory motifs;
2. Plasmids containing constitutive promoters, such as CMV (cytomegalovirus) or SV40 (simian virus 40) promoters;
3. Plasmids with cell type-specific promoters;
4. Plasmids with regulatable promoters.

(It should be emphasized once again that the plasmids listed above are mammalian expression plasmids; if the cells that one is working with are *Drosophila* or plant cells, for example, it will be necessary to refer to the counterparts of *minimal* and *constitutive* promoters for the respective organism.)

In this chapter, we will instead focus on a different group of vectors that are used as **reporters**. These vectors usually contain what is known as a **reporter gene** that is used to monitor the expression from plasmids transfected into cells of interest, and are the positive controls that cells are actually transfected, or they are used to monitor the activity of a promoter in a cell culture system. Traditional reporters include enzymes such as β-galactosidase (lacZ), luciferase, or fluorescent proteins such as green fluorescent protein (GFP) or its derivatives.

β-galactosidase (lacZ) has been studied quite extensively, and has therefore become the traditional molecule for many reporter assays. This enzyme has been mostly studied with respect to its role within the *lac* operon in bacteria, where it catalyzes the hydrolysis of lactose into its glucose and galactose monomers. The enzyme can also catalyze hydrolysis of other analogous substrates, most notably the colorless X-gal compound (5-bromo-4-chloro-indolyl-β-D-galactopyranoside), which gets hydrolysed into galactose and 5-bromo-4-chloro-3-hydroxyindole, which upon aggregation and oxidation forms an insoluble blue-colored complex (Figure 7.4a). This color reaction makes lacZ a powerful reporter in many applications. LacZ enzyme can also be used in complementation-based

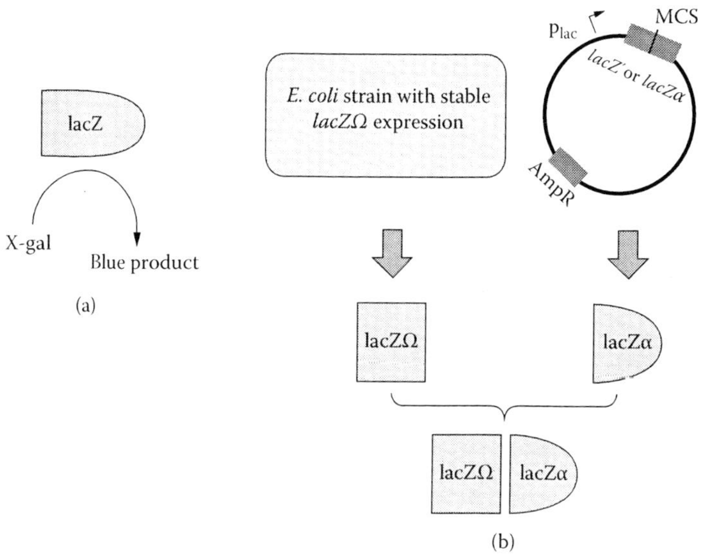

Figure 7.4 Schematic representation of how *lacZ* is used as a reporter. (a) A simplified scheme of the catalytic reporter activity of *lacZ,* whereby a colorless substrate X-gal is cleaved to generate a blue-colored product (see the text for details). (b) *lacZ* complementation is based on the principle that the enzyme can be split into two polypeptides that cannot function individually, but only exhibit catalytic activity when expressed in the same cell at the same time.

assays, where the *lacZ* coding sequence is split so that the enzyme gets translated as two separate polypeptides called lacZα and lacZΩ, which cannot exhibit any catalytic activity separately, but can show wild-type function when present at the same time (Figure 7.4b). In such systems, *lacZα* is usually provided in the plasmid DNA, and *lacZΩ* is stably integrated in the target cell, for instance, the bacterial strain.

Luciferase is an enzyme which was originally discovered in a different species of fireflies that can catalyze the oxidation of the substrate, luciferin, in the presence of ATP, producing oxyluciferin and bioluminescence. This has become a commonplace technique for observing promoter activities, particularly since the amount of bioluminescence emitted can be directly quantified with the help of a luminometer, this intentisty of light being directly proportional to the amount of enzyme synthesized in the cell (hence, the strength of promoter driving the expression of this enzyme).

Green fluorescent protein, GFP, has been discussed previously (Section 4.8.4 in Chapter 4), and some generic expression vectors have been given (Figures 4.9 and 4.10). However, in those cases GFP has been commonly used as a marker for expression, whereas in this section GFP will be used as a reporter to monitor how strong a regulatory element such as a promoter or enhancer is.

In all these cases, the reporter gene in question is present in either a promoterless vector, and the promoter of interest is cloned into the multiple cloning site, MCS (Figure 7.5), or the reporter gene is present under a minimal promoter (usually the minimal TK promoter), and the enhancer or regulatory elements are cloned into the MCS.

Depending on the choice of the reporter (which is related to the experimental system and the cell type used), the appropriate method of detection is used to monitor how many of the reporter genes are expressed, which can be used as a direct correlate of how strong the regulatory DNA sequence is.

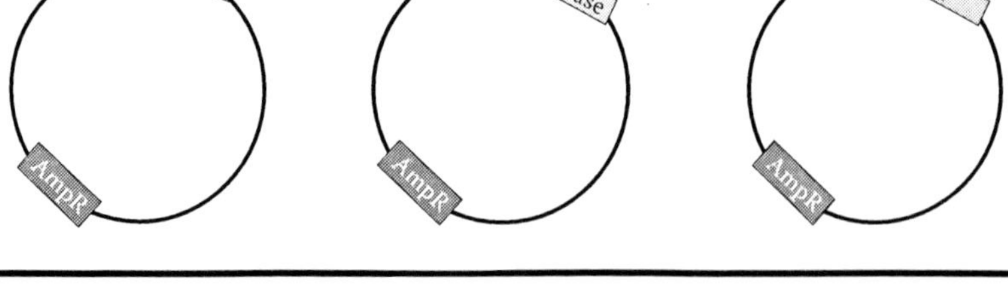

Figure 7.5 Simplified diagrams of typical reporter vector backbones. Hypothetical LacZ, luciferase, and GFP reporter vectors are shown (left to right).

7.3 Types of Transfection

Transfection in general refers to the introduction of exogenous DNA into cells; however, this DNA does not necessarily get integrated into the genome and may remain extrachromosomal. Thus, based on whether there is stable integration or not, transfections can be broadly categorized as transient and stable transfections.

7.3.1 Transient Transfection

Transient transfection refers to cases where the gene transferred to the cell is not integrated into the genome of the cell, and therefore gets diluted to the daughter cells with each cell division, and in the absence of selection pressure. Therefore, a high level of expression from the transfected gene only lasts for a few days after transfection, and then fades away as the DNA gets lost. Usually, high-copy number vectors that contain strong promoters are used for the high-level expression of proteins, markers, or reporters, followed by the desired analyses. It is a rather fast analysis, however, it bears the disadvantage that (on top of not being integrated and thus becoming diluted over time), the plasmid DNA is present in large numbers per cell, and thus creates a discrepancy with the rest of the genome that is diploid. The results of transient transfections, therefore, have to be carefully interpreted, since the data obtained could well be the result of overexpression of proteins.

7.3.2 Stable Transfection

In stable transfection, unlike the transient one, the transfected DNA eventually becomes integrated into the genome, albeit with varying probability (the efficiency of integration depends on cell type, method used, or selection pressure applied). However, the results of stable transfections are more reliable in the sense that there is eventually only one copy of the transfected gene, rather than tens or hundreds, and this copy is therefore more physiological, on top of which the stable integration ensures transmission to daughter cells over many generations. Viral vectors are usually preferred for stable cell line generation since they have the capacity to integrate into the genome of their target cells.

7.3.3 Recombination and Integration into the Genome

Recombination is a mechanism through which DNA from different sources (different organisms, cells, or chromosomes) can be broken and joined. Naturally occurring recombination events include homologous recombination, such as in crossing over; site-specific recombination, such as in integration of the viral genome; and nonhomologous end-joining, as in a DNA double-strand break repair pathway. Many of the key enzymes involved in recombination events may belong to the same family, such as Rec recombinase enzymes. These recombination principles are exploited in many of the genetic tools for the manipulation of animals or plants, therefore we will take a quick look at the basics of homologous and site-specific recombination events (Watson et al. 2008).

7.3.3.1 Homologous Recombination

Homologous recombination is a process in which DNA molecules with overall *similarity* (not *identity*) exchange corresponding parts of their sequences. It is mostly studied in crossing over during meiosis, and in post-replicative recombination repair. Homologous recombination can occur between any two of the recombination hot spots distributed almost randomly throughout the sequence in each chromosome—there is no single *site* from where the recombination originates.

The strand exchange is initiated with a double-strand break in one of the chromosomes, and short stretches of 3′ overhangs are created by trimming some nucleotides in these double-strand break sites. These 3′ overhangs, now single strands, bind to a single-strand binding protein, RPA, invading similar sites in the other DNA molecule (hence *homologous* recombination). A four-way Holliday junction forms at the site of exchange (this Holliday junction structure is also observed in site-specific recombination, although the context is slightly different there). Resolution of the Holliday junction (which by itself is an entirely complex process that will not be discussed here) generates DNA molecules that have exchanged *strands* (Watson et al. 2008).

Such a recombination mechanism is also used in double-strand break repairs known as the *post-replicative recombination repair*, where DNA damage that cannot be repaired is skipped during replication, leaving the original damage in place.

7.3.3.2 Site-Specific Recombination

Site-specific recombination, in principle, relies on very similar mechanisms to homologous recombination, however, recombination occurs at a specific site of around 30–200 bp, outside which very little or no homology is required. These sites are typically asymmetric in nature, enabling the recombinase enzyme to recognize the left and right *halves* of this motif. The recombinases that are mostly used in genetic modification are the Cre and Flp recombinases (which will be extensively covered in the following chapters). Strand exchange again takes place through Holliday junctions (Watson et al. 2008). The strand exchange will have different outcomes, such as integration, deletion, or inversion, depending on the source or strands to be exchanged (same or different DNAs), and the orientation of the recombination motifs (Figure 7.6).

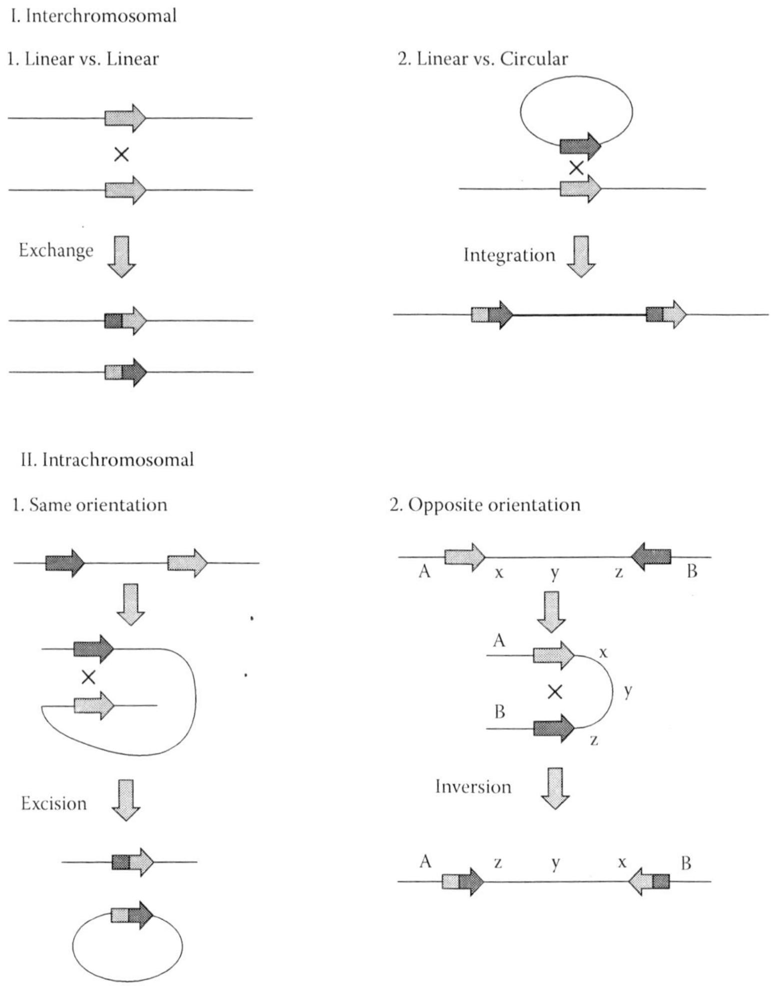

Figure 7.6 A summary of possible products of site-specific recombination. I. Interchromosomal exchange (or intermolecular exchange) can occur between either (1) two linear DNA molecules, which result in a strand exchange from a specific recombination motif (shown with two boxes and a cross), or (2) one linear and one circular DNA molecule, which results in the integration of the circular DNA to the linear one. II. Intrachromosomal (or intramolecular) exchange occurs within the same DNA molecule; however, in this case the recombination motif can either (1) be in the same orientation, which results in an excision of the region in the middle as circular DNA, or (2) in the opposite orientation, which results simply in the inversion of the middle region.

7.4 Level of Expression

The other feature to be considered is when the expression exogenous genes in the cells are the level of expression, which directly correlates with the promoter that is used in the vector. The expression of the gene of interest exploits the same mechanisms used in the gene expression in cells: a strong promoter will result in a higher level of gene expression, whereas a weaker promoter will cause fewer transcripts to be made from the gene in question. Some promoters will be *on* in every cell type, under any condition, whereas others will be either tissue-specific, or developmental time-specific, or expressed only under certain conditions in the presence of inducing agents. Thus, these very same principles will be utilized in the genetic manipulation of cells.

7.4.1 Constitutive Expression

Constitutive expression refers to the cases where the gene that is transferred into cells will be expressed at all times, under all conditions. Many mammalian expression vectors such as the pCMV series (Figure 2.21 in Chapter 2) use strong promoters that are always active. This is particularly preferred if high amounts of protein expression are required for preparative purposes.

7.4.2 Inducible Expression

Inducible expression in cells is achieved by selecting a promoter that is regulated by a variety of stimuli, which in turn regulates the activity of transcription factors. Some promoters are activated by heat stimulus (e.g., heat shock promoters); some are activated by chemicals (e.g., lactose-, lactose analog IPTG-, or tetracycline-inducible promoters); some others are regulated by the presence or absence of oxygen (such as hypoxia-inducible promoters).

Tetracycline is perhaps one of the most common inducing agents used in mammalian inducible expression systems, therefore, here we choose to explain the basic principles

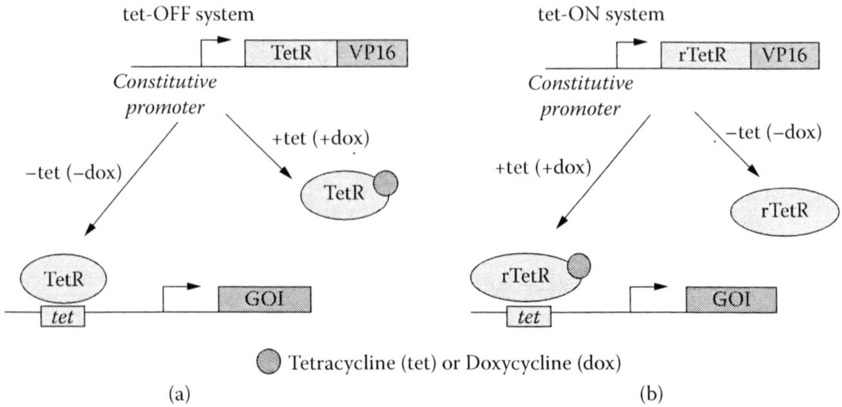

Figure 7.7 A schematic diagram of a tetracycline-inducible system. (a) A tet-OFF system relies on the binding of the tetracycline repressor (TetR) to the *tet* motif on target promoters. In this system, TetR is fused to the acidic activation domain of the VP16 transcription factor, and positively regulates transcription from the target promoter in the absence of an inducer, upregulating the expression of the gene of interest (GOI). TetR itself is released from the *tet* motif upon binding of tetracycline (tet) or its analog doxycycline (dox) to the TetR, thereby releasing the TetR-VP16 fusion from the inducible promoter, hence, tet-OFF. (b) In the tet-ON system, TetR is engineered such that its DNA-binding properties are reversed: rTetR binds to the promoter in the presence of tet (or dox), and is released from the promoter in the absence of the inducer. Therefore, the GOI is turned ON in the presence of tet (or dox).

of tet-ON and tet-OFF systems as examples. Essentially, in **tet-ON** systems the transfected gene is turned *on* in the presence of tetracycline, or its analog doxycline, in a dose-dependent manner (i.e., the higher the tetracycline, the higher the expression level); whereas in **tet-OFF** systems the transfected gene is turned *off* in the presence of tetracycline or doxycline (see Figure 7.7 for an overview).

7.5 Problem Session

Q1. What is the difference between a tissue culture, organ culture, and cell culture? If you were to design an experiment working with blood lymphocytes, what kind of culture conditions would you need?

Q2. You wish to study how the USF transcription factor regulates the *mitf* promoter sequence in response to UV, and through promoter deletions you wish to analyze whether the E box motifs on the *mitf* promoter are directly involved in this regulation. Would you choose to carry out this analysis in a melanoma *cell line* as a model, or use *primary* melanocytes instead, and why?

Q3. In the question above, you need to genetically modify the model cells using viral vectors (you want to integrate your target construct to the cells' genome). Which viral vector would you choose if you were working with a melanoma cell line? Which vector would be more appropriate if you were to work with primary melanocytes?

Q4. You want to modify an embryonic stem cell with your gene of interest, so as to generate a transgenic animal afterward. But the gene you are interested in expressing in the embryo may in fact be embryonic lethal. What kind of a strategy would you employ to overcome this obstacle?

Q5. You have the coding sequence for a pancreatic β-islet-specific gene sequence below, and you wish to transfer this gene into primary adipocytes so as to try and reprogram the adipocytes with this single gene. (a) How would you carry out the cloning? (b) How would you generate the viruses and transduce the adipocytes?

5′-ATG CCT CTC GTA GGG AGA TCC GAA CAG CAG TGA –3′

> Research is what I'm doing when I don't know what I'm doing.
>
> **Wernher von Braun**

Chapter 8

Genetic Manipulation of Stem Cells and Animals

> In the beginning there is the stem cell; it is the origin of an organism's life. It is a single cell that can give rise to progeny that differentiate into any of the specialized cells of embryonic or adult tissue.
>
> **Stewart Sell**
> *Stem Cells Handbook*

In the previous chapters, we learned about some basic terminology and principles of manipulation. This chapter will essentially apply this information to the genetic manipulation of stem cells and animals; and in the next chapter we will apply similar technologies to the manipulation of plants.

Stem cells have been known for decades, and they have been used for knockout animal production or transgenics for more than a score of years. Mario Capecchi, Sir Martin Evans, and Oliver Smithies were awarded the Nobel Prize for Physiology or Medicine in 2007 for their discoveries of the "principles for introducing specific gene modifications in mice by the use of embryonic stem cells," which they had been working on since the late 1980s. And in 2012, the Nobel Prize for Physiology or Medicine was awarded to Sir John Gurdon and Shinya Yamanaka "for the discovery that mature cells can be reprogrammed to become pluripotent" (see The Official Web Site for the Nobel Prize at http://www.nobelprize.org/nobel_prizes/medicine/laureates/2012/press.html). So the excitement around stem cells in the late 1990s until now was not for their identification, but rather for the potential of genetically manipulating and thereafter using them for other purposes, medical or commercial.

The term *stem cell* was actually coined back in 1908 by the Russian histologist Alexander Maximov to describe what we today refer to as *hematopoietic stem cells.* In the 1960s, Joseph Altman and Gopal Das presented the scientific evidence of adult neurogenesis (and thus the existence of adult neural stem cells), and in 1992 neural stem cells were cultured *in vitro.* In 1963, the presence of stem cells was shown in the bone marrow, and the presence of hematopoietic stem cells in human cord blood was discovered in 1978. What was exciting was that the first human stem cell line was established in 1998 by Thomson and his team, and from 2000 onward, the application of stem cell technology was on the rise. In 2001, Advanced Cell Technology cloned the first early human embryos for the purpose of generating human embryonic stem cells; and in 2003, adult stem cells were discovered in children's primary teeth, which was a relief in the midst of heated discussions on the ethics of using embryonic stem cells. In 2006, pluripotent cells were induced from adult cells (induced pluripotent stem, iPS, cells); and later repeated for the reprogramming of mouse skin cells and of human fibroblasts in 2007.

8.1 Stem Cell Technology and Knockout Cells

The widely accepted standard definition of stem cells relies on two major properties: (1) **self-renewal**, that is, the capacity to generate more stem cells like itself, and (2) **potency**, that is, the capacity to give rise to various differentiated cell types.

The ultimate *stem cell* in that respect is the fertilized egg, the zygote, which gives rise to an entire embryo with all the necessary extra-embryonic tissues necessary for its survival, which therefore is called **totipotent** (total potential). What are known as *embryonic stem cells,* or ES cells, however, are typically isolated from the inner cell mass of the blastocyst and are merely **pluripotent**, meaning they can give rise to many different cell types of the embryo, but cannot produce any extra-embryonic tissues such as the placenta. From gastrulation onward, the embryo already has begun to differentiate, forming initially the three embryonic layers, ectoderm, mesoderm, and endoderm, and thus the capacity of these cells is already restricted—these cells progressively become more and more restricted in their respective lineages, going from **multipotent** to **oligopotent**, **bipotential** or **unipotential** precursors, sometimes also referred to as stem cells.

When stem cell manipulation is discussed, the source of stem cells becomes important. For genetic manipulations, researchers either use embryonic stem cells or ES-like cells such as embryonic germ (EG) cells, embryonic carcinoma (EC) cells, fetal stem cells, umbilical stem cells, or induced pluripotent stem (iPS) cells.

Human ES cells can be obtained from the inner cell mass (ICM) of blastocysts from surplus IVF (*in vitro* fertilization) embryos that are donated for research. Alternatively, pluripotent ES-like cells can be obtained from terminated pregnancies. Both of these procedures have been banned in many countries due to ethical considerations.

More ethical methods for obtaining human ES-like cells include the isolation of fetal stem cells from the amniotic fluid, the umbilical cord blood, or the umbilical cord itself (this latter is by far the hardest in terms of isolation). The efficiency of these methods, however, is not great. Alternatively, in animal studies, primordial germ cells are isolated from the embryo and used as pluripotent cells.

These cells can be either used to explore the molecular mechanisms of development, of how cells regulate their genetic programs so as to differentiate into various types of specialized cells, or to study the effects of drugs and drug candidates on cells, or used for clinical applications. Alternatively, stem cells can be used to generate transgenic organisms either to knock out genes and create disease models, or to knock in genes to study their effects.

For clinical use, that is, the transplantation of stem cells to patients, the general transplantation terminologies apply: stem cells to be transplanted can be (a) ***autologous***, meaning the patient's own stem cells are removed, stored, manipulated if necessary, and given back to the same person; (b) ***allogeneic***, meaning the cells are isolated from a genetically nonidentical person, stored, manipulated if necessary, and transplanted to the patient; (c) ***syngeneic***, meaning the stem cells to be stored, manipulated, and used are from a genetically identical or else immunologically compatible person, such as a relative of the patient; or (d) ***xenogeneic***, meaning the stem cells to be used are from an immunologically compatible as possible but nonhuman species (such as primates or pigs).

8.1.1 Genetic Manipulation of Embryonic Stem Cells

Since embryonic stem (ES) cells can be expanded to form embyroid spheres, or clones, they can be manipulated to stably express exogenous genes, introduced either by viral vectors, by electroporation, micromanipulation, or transfection (see Chapter 7). Random integration (or insertion) is commonly used to overexpress or mutate genes for large-scale screening purposes or to integrate reporter genes for monitoring differentiation or other developmental events; gene targeting in ES cells is employed to exchange endogenous genes with engineered ones (either to knock out a functional gene by a homologous recombination with a nonfunctional version, or vice versa). There are various techniques employed, and more advanced techniques appear every day, but since this is an undergraduate level textbook we will concentrate on the main principles using some of the basic technologies.

The most crucial aspect of this study is no doubt the culturing of ES cells in an undifferentiated state, and differentiating them to the desired cell type when required. Human ES (hES) cell culturing comes with a multitude of ethical issues and hence is not possible in many countries, therefore we will mainly concentrate on the mouse ES (mES) cells and only briefly mention hES culturing.

In order to ensure that ES cell cultures remain undifferentiated upon propagation *in vitro*, one must monitor the maintenance of some common features, such as (1) the expression of ES cell markers (e.g., Oct4, Nanog, SSEA1, etc., that are common to both mES and hES cells; for more detailed information on the characterization of hES cell lines, please refer to the *Nature* biotechnology article by The International Stem Cell Initiative 2007); (2) forming teratomas when implanted to animals; and (3) the ability to produce cell types from all three embryonic germ layers. ES cell cultures are also routinely monitored for any chromosomal abnormalities. Therefore, when manipulating ES cells genetically, similar strategies as discussed in Chapter 7 will be used, but care must be taken to keep them undifferentiated until necessary.

When differentiation into a specific cell type is wanted, a number of different strategies are employed. The normal differentiation pathway for an embryo can be used; for example, erythrocytes are differentiated from multipotent hematopoietic stem cells (HSCs), which can be isolated from organisms as adult stem cells (ASCs) (for example, from the bone marrow) or obtained from ES cells through a combination of an HSC-specific differentiation medium. HSCs are normally found in the bone marrow and give rise to cells of the myeloid and lymphoid lineage, such as erythrocytes, neutrophils, megakaryocytes, T-cells, and B-cells.

Just like bone marrow transplants, obtaining HSCs from bone marrow is also a caveat, no matter how advanced the technology has become. Therefore, obtaining HSCs from ES cells is a cherished alternative. However, in spite of the beautifully simple strategies outlined above, the first attempts have not proven extremely successful, with Bhatia's group achieving 1% of the ES-derived HSCs successfully engrafted to immunocompromised mice (Wang, Menendez et al. 2005). A more recent work has improved the protocol by growing and differentiating the ES cells on stromal cells obtained from the area, thereby providing the ES cells with the appropriate environment, or niche (called a *stem cell niche*). But researchers have also shown that transforming growth factor-β (TGF-β) was almost equally successful for efficient differentiation—2 to 16% functionally engrafted HSCs (Ledran et al. 2008). These studies required no genetic modification.

In a different study, genetic modification of mES cells with tetracycline-inducible Cdx4 has resulted in successful differentiation into HSCs and successful engraftment to irradiated mice (Wang, Yates et al. 2005). A more recent study has combined conditional (tetracycline inducible, tet-OFF system) HoxB4 expression in mES cells with coculturing on stromal cells (Matsumoto et al. 2009).

Genetic modification methods for ES cells are constantly being improved (for an example, see Braam et al. 2008), but they are essentially not much different from that of mammalian cells—retroviral or lentiviral vectors are commonly used by many researchers. (Electroporation is not the desired method due to low rates of survival, and lipofection does not yield high enough transfection efficiencies.) When using retroviral vectors for ES cell manipulation, however, it was observed that sustained gene expression from integrated provirus was difficult due to high levels of *de novo* methylation in ES cells, thereby promoting repression from the transgene.

Lentiviral vectors offered significant advantages for ES cell manipulation, particularly because expression was not silenced in the undifferentiated as well as differentiated stem cell, and transduction efficiency was higher than that of retroviral transduction.

Expression of the transgene from a tissue specific promoter is the method of choice when expression from this gene is only desired in one type of cell after ES cell differentiation. For example, the myosin light chain promoter is used for expression of the transgene only in myocytes upon differentiation. This method is particularly used for random integration.

However, homologous recombination can be exploited when the transgene is to be targeted to the location of its endogenous counterpart, especially for knockout or gene therapy applications. In 1987, Thomas and Capecchi showed how transgenes could integrate to the mouse ES genome through homologous recombination. Essentially, the homologous recombination targeting vector contains regions homologous to the targeted gene, between which a marker is inserted that will disrupt the endogenous gene (Figure 8.1).

8.1.2 Induced Pluripotent Stem Cells (iPSCs)

Because of the ethical issues surrounding the use of human ES cells for clinical purposes, the need for more ethically appropriate as well as economically feasible solutions arose. The ES cells that were previously unmatched in their capacity to generate a variety of cell types soon became replaceable with iPS cells, which in effect are dedifferentiated adult somatic cells.

The study of ES cells has generated a number of clues about the so-called stemness genes, which were tested for their abilities to convert differentiated somatic cells into *embryonic-like* cells (Bhattacharya et al. 2004; Li and Akashi 2003; Okita, Ichisaka, and Yamanaka 2007; Richards et al. 2004; Takahashi and Yamanaka 2006). These dedifferentated or reprogrammed cells could be obtained from a wide range of somatic tissues,

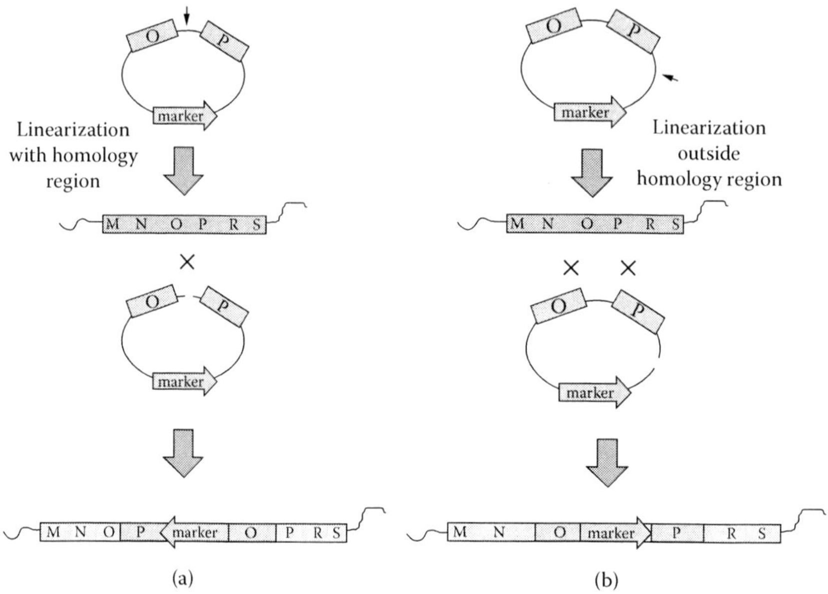

Figure 8.1 Schematic diagram of the basis of homologous recombination used for transgene insertion. (a) If the vector is linearized within the region of homology, then the marker is inserted into the target region in the opposite orientation. (b) If the vector is linearized outside the region of homology, then the marker is inserted into the target region in the same orientation.

including epidermis, mesenchyme, dental mesenchyme, and so on, making them much more acceptable in ethical terms. Human iPSCs thus generated were found to express many stem cell markers and could differentiate into cells of all three germ layers, with significant clinical implications.

Reprogramming can be achieved by a transfection of the somatic cells by a number of reprogramming genes, such as c-Myc, Sox2, Oct3/4, and Klf4; however, different researchers have optimized their own variations on this general recipe (Okita, Ichisaka, and Yamanaka 2007; Takahashi and Yamanaka 2006). C-Myc is a proto-oncogene, which resulted in an increased incidence of cancer in reprogrammed cells, therefore a cocktail without c-Myc later came to be used for iPSC generation by the Yamanaka team (Nakagawa et al. 2008), while in a different study, loss of p53 (a tumor suppressor) was found to enhance reprogramming efficiency of cells (Kawamura et al. 2009).

Viruses and, in particular, lentiviral vectors, have been used extensively thus far for the reprogramming of iPSCs, however, nonviral transfer methods are currently under investigation. Regardless of the transfer method, however, the efficiency of reprogramming somatic cells into pluripotent cells is extremely low, around 1 to 5%, making it difficult for routine clinical applications at present.

8.2 Transgenic Animals

A transgenic animal is defined as an animal carrying a foreign gene that is inserted into its genome. This genetic modification is usually done at the ES cell stage, or at the pronucleus stage, although other methods are also available. See Figure 8.2 for an example of transgenic mouse generation, an explanation of the illustration is as follows: The first step is to insert the loxP sites to either side of the target gene (which is embryonic lethal). In order to do this, homologous recombination can be used to replace the endogenous gene with a gene flanked with loxP sites (see panel on the left). This vector can either be introduced to the male pronucleus of the fertilized egg, or the embryonic stem (ES) cells, and the embryo that is thus generated is transferred to a foster mother from a different mouse strain (middle panel). The chimeric mice born to the foster mother are further crossed to the mouse of the same strain as the foster mother, until a pure heterozygous transgenic mouse of the original donor strain is obtained. This transgenic mouse still has the gene and expresses it, therefore it can survive the embryonic period. If this transgenic animal is then crossed to a second transgenic that expresses

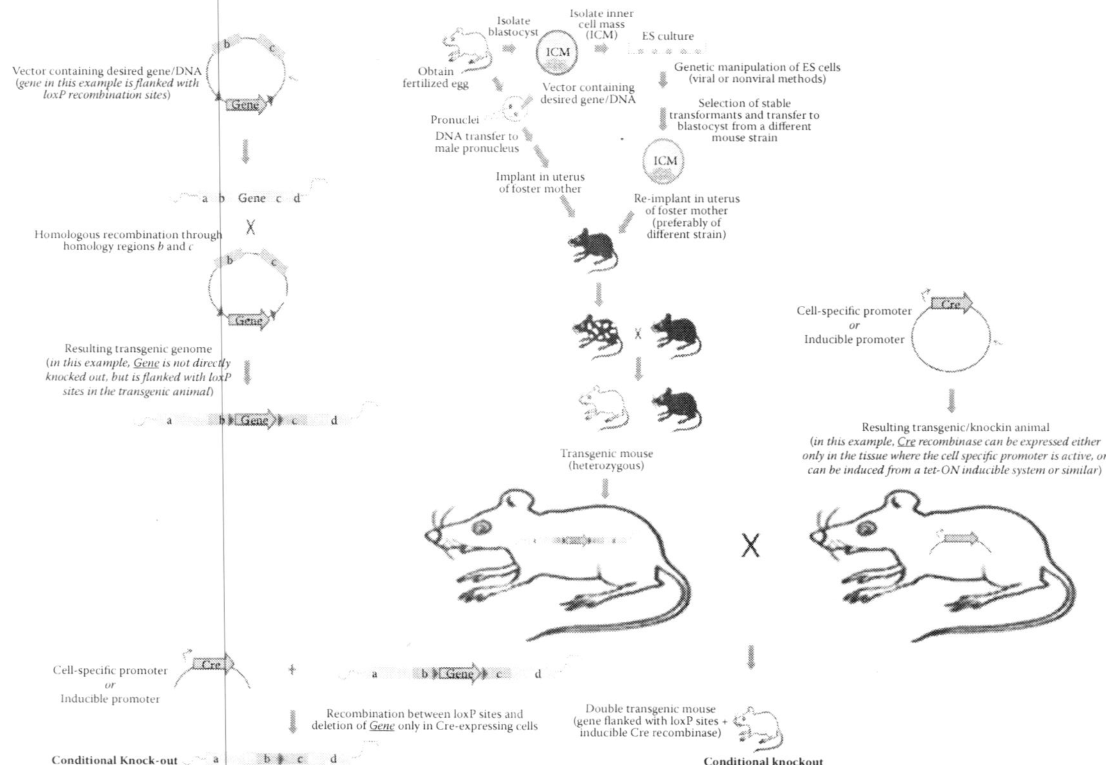

Figure 8.2 Schematic diagram of conditional knockout generation. (See the text for an explanation.)

Cre recombinase either from a cell-specific promoter or from an inducible promoter (right panel), the double transgenic obtained from this cross can be considered a conditional knockout: Cre recombinase will cause site-specific recombination between loxP sites and cause removal of the gene from the genome either in a specific cell type (if Cre is expressed from a cell-specific promoter), or only upon induction by an agent at a desired embryonic or fetal or postnatal period (if Cre is expressed from an inducible promoter).

Transgenic technologies have been in development since the early 1980s; transgenic mice with growth hormone phenotypes (due to transgenic human growth hormone expression, called a *supermouse*) were reported in 1982 by Palmiter and friends (Palmiter et al. 1982; Palmiter and Brinster 1986). This was followed by transgenic chickens, pigs, sheep, rabbits, fish, goats, and sheep.

The first generation transgenic animal is commonly called the *transgenic founder*, meaning that the transgene is not necessarily transferred to the next generations (i.e., is not inheritable). Only when this transgenic animal is bred and the transgene is shown to be stably heritable over generations is it called a *transgenic line*.

Transgenic animals have been generated for a number of different applications, mostly using animals as *animal bioreactors* to produce a number of pharmaceutical (known also as *animal pharming*) or otherwise industrial products. Most of the time, the milk of goats, sheep, and so on, are used as a renewable medium for the production of functional enzymes, antibodies, or other proteins, exploiting the existing dairy production technologies. To achieve this, the transgene is placed downstream of the promoters of mammary-specific genes. For instance, α-1-antitrypsin, which is used in the treatment of cystic fibrosis, is expressed from a β-lactoglobin promoter in sheep; antithrombin III, which is used in open heart surgeries, is expressed from a cow casein promoter in goats; and lactoferrin, which is used as an infant formula additive, is expressed from a cow casein promoter in cows (Castro et al. 2010).

Similarly, transgenic chickens can be used to express functional proteins in the egg white from a tissue-specific promoter, which could then be extracted using existing process techniques. The method of transgenic animal production has to be optimized to the organism in question; however, the basic logic is very similar to producing a transgenic mouse, in the sense that either the foreign gene to be knocked in is expressed from a cell type-specific promoter, or an endogenous gene is replaced with a genetically engineered version through homologous recombination, and so forth, although somatic cell nuclear transfer is the preferred method in some laboratories (see Section 8.3).

In the cloned pharm animals Molly and Polly, generated by the Roslin Institute in Scotland, for instance, Factor IX is expressed from the bovine β-lactoglobin promoter that is active in sheep mammary gland cells, and is co-transfected with a vector carrying the neomycin resistance gene as a selection marker to diploid fetal cells isolated from an impregnated sheep. The genetically modified stable cells that have thus been obtained are then fused to enucleated eggs isolated from a Scottish ewe, and the resulting blastocyst is implanted to a recipient ewe, giving birth to the transgenic pharm animals (Colman 1999; Wilmut, Sullivan, and Taylor 2009).

Knockout (KO) technology exchanges the wild-type and functional allele of a gene with an inactive one, thereby effectively eliminating or inactivating a gene. This technology is particularly useful when creating disease models, however, if the gene that is inactivated is somehow crucial for embryonic development, then deleting or inactivating this gene would be *embryonic lethal*. If that is the case, then an inducible promoter is chosen to express the inactive version of the gene (Figure 7.7), or to drive the expression of the recombinase enzyme (in combination with specific recombination sites). This is called a **conditional knockout**.

Site-specific recombination is another method that can be exploited for inducible integration of a transgene into the genome: usually a recombination site such as loxP or frt is introduced into either end of the transgene. Co-expression of a Cre recombinase (from bacteriophage P1, which recognizes the loxP sites) or Flp recombinase (which recognizes frt sites), under an inducible or tissue-specific promoter as necessary, catalyzes the conditional recombination between these specific sites at the desired embryonic time or the desired tissue (Figure 8.2).

BAC genomic libraries have also been used in association with transgenic animal models, since they were found to reduce positional effects of transgenes, and since transgenes were expressed at physiological levels and as such exhibited developmental timing and expression patterns as in the source organism (Beil et al. 2012; Van Keuren et al. 2009) (see Chapter 2, Section 2.2.4 for BAC vectors). The most effective and preferred mode of delivering a BAC library or construct into an animal is pronuclear microinjection due to the rather large size of constructs, although viral vectors or electroporation to ES cells have also been reported (Van Keuren et al. 2012). The microinjection method was also observed to be independent of construct size and form (linear or circular), whereas the BAC DNA concentration injected into fertilized eggs appeared to be more important (with high DNA doses exhibiting higher toxicity to the eggs) (Van Keuren et al. 2012).

Although BAC transgenesis has largely replaced conventional transgenesis methods due to its above-mentioned advantages, it also comes with its own set of disadvantages. BAC transgenesis occurs through random integration, and while positional effects are reduced, the number of copies can vary. The relatively low efficiency of generating transgenic founders means that more pronuclei have to be injected with BAC constructs to achieve the same number of transgenic animals. And constructing a BAC library might take weeks to months, and when combined with the time required to inject more pronuclei (which may take much longer than in traditional transgenesis methods) the setup can be more time consuming (Beil et al. 2012).

Gene trapping is a random knockin that is essentially used to randomly interfere with different genes' expression so as to identify genes associated with a particular phenotype, in large-scale and high-throughput genetic screens. The gene trap vector can be of two versions, as discussed next, but ultimately it is designed to prevent RNA splicing of genes into which the transgene is inserted.

In one kind of gene trap vector, the trap cassette is flanked with a promoter upstream and a splice donor site downstream; in the other, the trap cassette is flanked with a splice acceptor upstream and a polyadenylation sequence downstream (Figure 8.3a). Either one or the other vector can be chosen, or the two vectors may be used in parallel for confirmation, but the idea is to create a large number of transgenic mice that have random integration of such a vector in their genome (Figure 8.3b). Since it is random integration, if the vector integrates to a chromosomal region devoid of any protein-coding gene, there will be no interference with gene expression, hence no phenotype will be observed. If, however, the vector integrates within a gene, depending on whether the first or the second type of vector was used, either one of two outcomes is possible:

1. Either the first type of vector integrates within a gene, and the trap cassette is expressed from the promoter and is spliced to exons downstream of the integration site, possibly also interfering with the expression of the trapped gene from its own promoter (see Figure 8.3c, right-hand side);
2. Or, the second type of vector integrates within a gene, and depending on the position of integration, the trap cassette is spliced to the closest upstream exon, however, transcription is stopped at the polyA signal of the vector, hence none of the downstream sequences of the trapped gene are expressed (see Figure 8.3c, left-hand side).

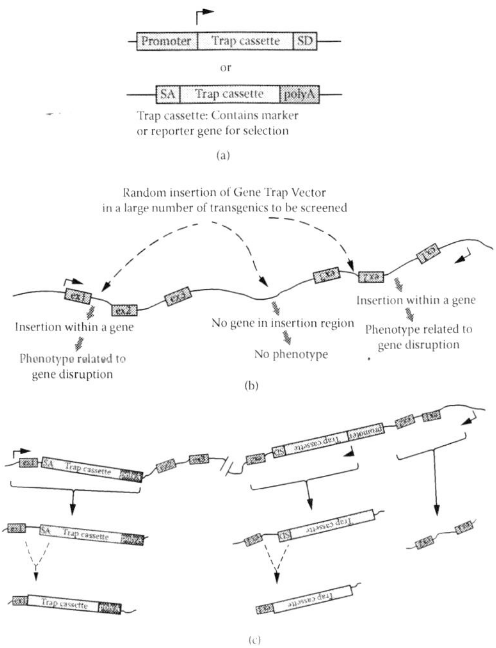

Figure 8.3 A schematic diagram of the gene trapping principle. (a) The two types of gene trap vectors. (b) Random integration of the gene trap vector to the genome of mice; large number of such random transgenics will be analyzed for the desired phenotype. If the gene trap vectors integrate within a gene, expression from the gene will be disrupted to various degrees of severity (see text for details). If, however, the vector is integrated in a chromosomal region that harbors no gene, then no gene expression will be detected, hence the desired phenotype will not be observed. (c) Two different types of gene trap vectors integrated to different gene regions are shown here to represent the functioning principle; if the first type of gene trap vector that contains an SA motif, a trap cassette, and a polyA signal is integrated within an intronic region, it will get transcribed into a primary transcript, and will be spliced into the mature mRNA through the SA site (left side). If the second type of gene trap vector that contains a promoter, a trap cassette, and an SD motif is integrated within an intronic region, it will interfere with the transcription of the trapped gene (ex1 and ex2 may or may not be transcribed properly; at best, the primary transcript will include the entire gene trap sequence), but also the trap cassette will be transcribed from its own promoter and spliced to ex3 of the trapped gene (right side). (SA: splice acceptor; SD: splice donor; polyA: polyadenylation signal; ex: exon.)

It should be noted that integration of the vector closer to the 5′ end of a gene region will yield a more complete inactivation of the gene and hence more severe phenotypes than integration closer to the end of a gene, thus yielding severe-to-mild phenotypes depending on the site of integration. The gene trap vectors can also be modified to contain recombination sites so as to generate *conditional gene traps*.

8.3 RNA Interference and MicroRNAs

As briefly mentioned in Chapter 4, Section 4.2, antisense RNA has long been used to inhibit protein synthesis, and hence to study the function of genes. In an interesting study performed by Andrew Fire and Craig Mello in 1998, however, an interesting phenomenon was consistently observed: the double-stranded RNA (sense and antisense together) that was supposed to be a *negative control* of the antisense RNA, started some twitching motion in the *Caenorhabditis elegans* muscle, more prominent than the antisense-injected worm (Fire et al. 1998). Fascinated (or perhaps frustrated) by this, Fire and Mello conducted a series of very elegant experiments and showed that the **RNA interference** mechanism was very effective in silencing gene expression, which led to their Nobel Prize (The Official Web Site of the Nobel Prize at http://www.nobelprize.org/nobel_prizes/medicine/laureates/2006/press.html). RNA interference was later shown to also exist in *Drosophila*, *C. elegans*, and mammalian systems. It then became a handy tool for studying gene expression, since it was much easier (took relatively less time) than generating knockout mice. Many vectors exist (more commonly lentiviral or other viral vectors for delivery) that transcribe a **short hairpin RNA (shRNA)** (Figure 8.4a).

RNA interference was initially identified through the action of **short interfering RNA (siRNA)** molecules, which are 21–23 nt-long dsRNA molecules (Figure 8.4b). Believed to be

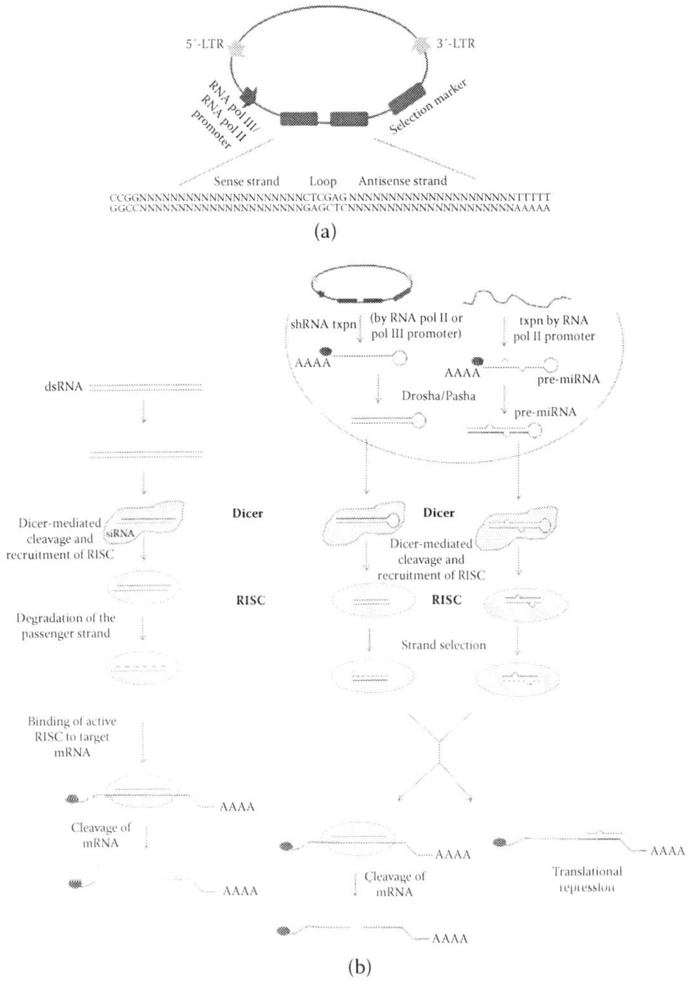

Figure 8.4 Major RNA interference mechanisms are discussed in the text. (a) A diagram of a typical shRNA-expressing plasmid. shRNA is typically transcribed from an RNA pol III promoter, such as a U6 promoter; however, more recently RNA pol II promoters are also being employed in vectors. The transcript contains complementary regions that will form the stem, and a loop stretch in between. (b) dsRNA, shRNA, and miRNA are all processed through similar mechanisms in the cytoplasm, cleaved into shorter fragments by Dicer, and loaded onto the RISC complex, where one of the strands is degraded, and the remaining strand binds to the target mRNA. This binding can either result in mRNA cleavage or translational repression, both of which will result in interference with gene expression.

originally a self-defense mechanism against double-stranded RNA genome parasites, such as viruses or transposons, the key enzyme in RNA interference, Dicer, cleaves dsRNA molecules into shorter siRNA fragments. The two strands of this siRNA are denatured into single strands, one of which (the passenger strand) is degraded and the other (guide strand) is incorporated into the other key component of this machinery, the RISC complex (RNA-induced silencing complex). The activated RISC complex then binds target mRNA, which results in the cleavage of the target (hence interference with gene expression) (Davidson and McCray 2011) (Figure 8.4b).

Later, small noncoding RNA species that are transcribed from RNA pol II promoters were discovered and named **microRNA** (or **miRNA**), due to their relatively small size. They were then found to be small, but developmentally extremely important. The miRNAs were processed pretty much through the same mechanism as dsRNAs, with the major difference that they are transcribed by RNA pol II in the nucleus as primary miRNA (pri-miRNA), and thereafter processed by the Drosha/Pasha complex into pre-miRNA. This pre-miRNA then gets translocated out to the cytoplasm, where the Dicer complex processes it into shorter miRNA, and this miRNA then gets loaded onto the RISC complex, where strand selection takes place. The remaining strand on the RISC complex then binds to target mRNA—if this binding has perfect complementarity, then the result is a cleavage of target mRNA. If, however, there is imperfect complementarity, then the result is a transcriptional repression (Figure 8.4b).

Either way, expression of the gene is silenced (either by translational repression, or by mRNA cleavage).

An **shRNA** sequence is typically through transcription from a plasmid or viral vector, usually from an RNA pol III promoter such as the U6 promoter, although more recent vectors also employ RNA pol II promoters (Figure 8.4a). The resulting hairpin RNA is processed in the same manner as a miRNA, and thus can affect gene expression either through mRNA degradation or translational repression. The shRNA-expressing vectors can also be used in association with transgenic technologies to create tissue-specific and/or inducible RNA interference in either animals or plants.

8.4 Animal Cloning

Cloning is essentially creating genetically identical copies, and in nature genetic clones are found in monozygotic twins in mammals, parthenogenetic progenies in species such as amphibians, and in nuclear transplantation.

The concept of cloning was in a way discovered by Spemann back in 1938, when he used a nuclear transfer technique to split a salamander embryo into two. But cloning as we know it today was first generated for mammals in 1997, when researchers at the Roslin Institute cloned Dolly, and thereafter with the cloning of Rhesus monkeys (1997), calves (1998), pigs (2001), and cats (2002), among others, that were reported.

Mammals can be cloned simply by the twinning technique, which is the splitting of a cell off from an embryo so as to generate a whole new one, in a way similar to the natural twinning process. In this case, usually an embryo is split into two half-embryos, transferred to a surrogate mother, and identical copies are born.

The use of **somatic cell nuclear transfer (SCNT)** is another method that is used to generate pluripotent stem cells. This is what the Roslin scientists used to clone Dolly. Essentially, a tissue biopsy is taken from the donor, cells are grown in tissue culture, and the nuclei are isolated, carrying the genetic information of the donor. These nuclei are then transferred to the enucleated eggs (i.e., the haploid nuclei of the eggs are removed) of the recipient surrogate mother; thus the egg is now a hybrid of the recipient egg's cytoplasm and the donor cell's nucleus with 2n chromosomes. These genetically altered eggs are then grown *in vitro*, some producing embryoid bodies, and these cloned embryos are then transferred to the surrogate mother, until babies are born. It should be noted first that in the original research that produced Dolly, only 1 in 273 nuclear transfers were successful, and second, the eggs that were created still carry the maternal mitochondria of the recipient, not the donor, and as a result are not exactly a clone of the donor cell.

From an ethical as well as technical perspective, one key question is whether all clones are "normal." Rates of viable and successful animal clones are extremely low, and the reasons for the low success rates are numerous: high rates of abortion at different stages of pregnancy were observed; in addition various abnormalities were observed in cloned animals, depending mostly on the tissue of origin for the nuclei that are transferred (cumulus cells of the cattle and Sertoli cells of the mouse, for example, proved to have a better yield in SCNT) (De Berardino 2001). **Epigenetics**, heritable changes that are *not* directly related to DNA sequence, is assumed to be yet another reason behind the low success rates.

8.5 Pharm Animals

Whether through cloning or through transgenics, commercialization and application of genetically modified animals, particularly for use in the pharmaceutical industry, has given way to the term **pharm animals**.

There may, in fact, be several different reasons for transgenic or cloned animal generation: (a) production of superior livestock (in terms of breeding, nutritional value, taste, or other), (b) production of human therapeutic proteins or other biopharmaceuticals, (c) use as an organ source in transplantations, (d) resurrection of beloved yet diseased or dead pet animals, or (e) resurrection of extinct species and/or recovery of endangered species (there are serious ethical debates around these last two issues).

Use of transgenic or cloned animals as bioreactors for the production of biopharmaceuticals has attracted a lot of attention due to the market value of these products. To solve this particular problem, the dairy industry provides a valuable tool for large-scale production and purification, hence much attention focuses on transgenic cattle for the production of such biopharmaceuticals in milk. Human lactoferrin is one such clinically valuable macromolecule, particularly in the treatment of infectious diseases. Van Berkel and colleagues have created transgenic cattle that produced human lactoferrin in milk from a bovine β-casein promoter (Van Berkel et al. 2002). In 2009, the Food and Drug Administration (FDA) approved the recombinant human antithrombin-α (A-Tyrn) produced in the milk of transgenic goats from GTC Biotherapeutics (Kling 2009).

In 2000, A/F Protein Inc. (Waltham, Massachusetts) developed a transgenic salmon that was engineered to grow faster with less food; and yet it was greatly debated whether this transgenic salmon could overgrow its natural counterparts (Niiler 2000). More recently, the milk quality of cows has been significantly improved through the transgenic expression of β- and κ-casein (Brophy et al. 2003). In order to improve meat quality, Lai and colleagues expressed ω-3 fatty acid in pigs through a transgene (Lai et al. 2006). Similar to the dairy industry, another industry that has been exploited for the production of recombinant nutritional supplements or biopharmaceuticals is the egg industry. Harvey and colleagues (2002), using an avian leukosis virus (ALV)-based vector, stably expressed an exogenous β-lactamase protein from a CMV promoter in a chicken egg.

8.6 Gene Therapy

Gene therapy studies, in fact, date back to the 1990s, with the first gene therapy applied to Ashanti DeSilva in the laboratory of Dr. W. French Anderson of the National Heart, Lung, and

Blood Institute. Ashanti, then a four-year-old, had a defect in the gene coding for the enzyme adenosine deaminase (ADA), which results in immune system deficiency. (Severe Combined Immunodeficiency syndrome, SCID, for short. For a brief timeline of gene therapy trials, visit the NIH history page at http://history.nih.gov/exhibits/genetics/sect4.htm#2.) The success of these first trials meant that gene therapy quickly became a hot topic to study.

However, it soon became clear that gene therapy was only applicable to single gene disorders, and was much more useful for disorders of the immune system, since blood cells were replenished at regular intervals in the body, so modifying the source cells would help produce new cells that could synthesize the functional protein or enzyme. This was a serious drawback; in addition, three children treated for X-linked SCID developed leukemia in 2005 (due to the vector being inserted near an oncogene in the patients' genomes).

Gene therapy strategies generally fall into two major categories:

1. *Direct gene therapy*, where the functional genes are directly transferred to the patient. (Most commonly this is an extra functional copy, although direct correction of the mutated gene is also under investigation, which will be discussed in genome editing approaches in Section 8.7.)
2. *Cell-based gene therapy*, where live cells (either the patient's somatic cells from target tissue, or stem cells) are themselves used as vehicles to deliver functional genes into the patient's body (cell-and-gene therapy) (Figure 8.5).

So, essentially the technology used in gene therapy is in principle no different than that used for transgenic or knockout animal studies. *In vitro* or *in vivo* modification of the patient's cells (somatic, adult stem cell, or iPS) is ultimately required. The gene therapy vectors that are most commonly used to transfer the therapeutic gene either directly to the

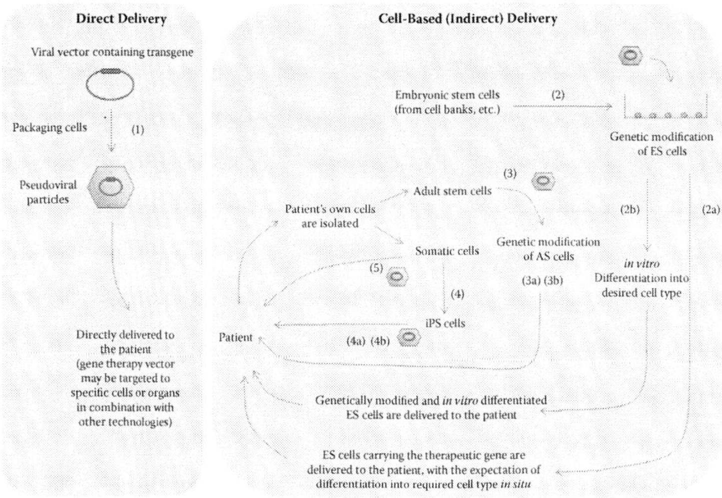

Figure 8.5 A brief outline of gene therapy strategies. A direct delivery strategy is based on constructing a virus-based vector to deliver the therapeutic gene (either as an extra functional copy integrated randomly to the genome, or to replace one of the defective genomic copies using homologous recombination) (left panel). This vector will be packaged into pseudoviral particles and delivered directly to the patient, commonly through injections (1). Cell-based delivery strategies are in general based on delivery of the therapeutic gene to cells outside the patient's body, and then delivery of these genetically modified cells back to the patient (right panel). There are several approaches, only some of which are summarized here. Embryonic stem (ES) cells (e.g., obtained from cell banks) are modified using gene therapy vectors (2), and are either transferred to the patient as genetically modified ES cells, expected to home in and differentiate into the desired cell type *in situ* (2a), or are differentiated into the desired cell type *in vitro* and delivered to the patient thereafter (2b). Alternatively, patient-derived adult stem (AS) cells may or may not be directly obtained from the patient, may be used for genetic modification by gene therapy vectors (3), and are delivered directly to the patient as genetically modified AS cells (3a), or are differentiated into the desired cell type *in vitro* and delivered to the patient thereafter (3b). Yet another approach preferred by some scientists is to use the patient's own somatic cells to generate iPS cells (4), and then proceed as in ES or AS cells (4a and 4b). One other alternative studied by some researchers is to directly transdifferentiate somatic cells into the desired cell type while "correcting" the genetic defect (5).

patient or indirectly through somatic or stem cells include retroviral, lentiviral, or adenoviral vectors, although nonviral delivery methods such as transposon-based and other vectors are also being studied (Sheridan 2011).

The first approved clinical application of direct delivery to Parkinson's disease patients was initiated in 2003 (Howard 2003), and later reported to give positive results (Sheridan 2007); however, there are still some drawbacks, and a routine treatment for Parkinson's is still far away. On the other hand, by employing cell-and-gene therapy principles, researchers have succeeded in showing proof-of-concept that human ES cells can effectively be differentiated into pancreatic β-islet cells and produce insulin, as a putative novel approach to treating diabetes (Madsen and Serup 2006). However, to date the approach has not been reliable enough to routinely use in the clinic. Recently, neural stem cells have been approved by the Gene Therapy Advisory Committee in the United Kingdom for ischemic stroke, and two other commercial stem cell products (hESC-derived oligodendrocyte precursor cells and a human neural stem cell line) were approved by the Swiss regulatory agency for brain disorders and spinal paralysis treatment (Mack 2011).

From a clinical perspective, however, in spite of all the recent advances and approvals of trials, there are still many challenges to overcome. First, random integration of the gene therapy vector into an off-target cell and/or an off-target genomic location (which may lead to tumorigenic responses, as in the 2005 trials, noted above) must be avoided. If stem cells are used, there is always a risk that the stem cells used as gene delivery vehicles will develop into teratocarcinomas, thus any undifferentiated ES or iPS cells should be removed prior to transfer back to the patient as a safety precaution (Mack 2011). Second, the patient's immune system is always an issue to consider, as transgenes themselves and/or the vectors used to deliver the transgene, are by their very nature "foreign" and might trigger an immune response. Similarly, if nonautologous stem cells are used, the cells used as vehicles to deliver the transgene may eventually be rejected by the patient's immune system (Sheridan 2011). Last, it is still unclear whether cell-based therapies, in particular, indeed affect and *repair* the problem, or simply alleviate at least some of the symptoms indirectly, through secretion of growth factors, for example (Mack 2011).

In spite of these challenges, gene therapy is still an attractive area of research, and the success stories together with ongoing advances in technology give hope to everyone working in the field. There is a light at the end of the tunnel.

8.7 Genome Editing

Genome editing is a very recent technology, which has boomed in the past two to three years, and is already patented (by the Broad Institute of MIT and Harvard) and on the market as a quick-and-easy method of altering organisms' genomes (with the first primate twins born in Kunming, China, after successful editing of their genome) (Editorial 2014a; Larson 2014; Rojahn 2014): it was quickly chosen as the Method of the Year 2011 (Editorial 2012).

Genome editing is a method whereby a piece of DNA is inserted, replaced, or removed from the organism's genome, with the help of genetically engineered nucleases and the host cell's own double-stranded break repair machinery. Homologous recombination takes place between two DNA molecules that harbor significant sequence homology (see Chapter 7, Section 7.3.3 for a brief overview). Double-stranded breaks stimulate homologous recombination machinery, which is the basis for genome editing approaches. The double-stranded breaks in this case are generally achieved by nucleases that were engineered to recognize specific target sequences.

There are currently four families of engineered nucleases used for genome editing purposes:

1. Zinc finger nucleases (ZFNs)
2. Transcription activator-like effector nucleases (TALENs)
3. The CRISPR/Cas system
4. Engineered meganucleases

Zinc finger nucleases (ZFNs) are based most typically on the Fok I restriction enzyme that is fused to a zinc finger DNA-binding domain engineered to target a specific DNA sequence.

The **TALENs** are similar to ZFNs. Each DNA-binding domain of TALENs can recognize a different single DNA base, hence a combination of different TALENs can in practice be used to target any specific sequence on the genome. The endonuclease activity again is through the Fok I restriction enzyme. TALENs have major advantages over ZFNs. First, off-target mutation rates are generally lower, and they can be designed to target virtually any genomic sequence.

The **CRISPR/Cas system**, also discussed above with respect to primate genome editing, has generated a lot of excitement in that this system can achieve a relatively higher mutation rate (here is the catch: the off-target mutation rates are also higher) and is relatively easy and cheap. It consists of a target-specific *guide* RNA, and a nontarget-specific nuclease. CRISPR stands for *clustered regularly interspaced short palindromic repeats*, and cas genes are *CRISPR-associated genes.*

Meganucleases of microorganisms have naturally long recognition sequences (>14 bp), and with protein engineering, various meganuclease variants have been generated to cover a large plethora of unique sequence combinations. In addition, meganucleases have been known to cause less toxicity in cells compared to ZFNs or TALENs.

While traditional gene targeting methods in embryonic stem cells have been shown to introduce genetic variations in roughly 1 in 106 cells, nuclease-mediated knockout or knockin trials were shown to significantly increase the targeting rate to around 1 in 100 to 1 in 2 cells (Wang et al. 2013). And while traditional targeting methods take 6 to 12 months, nuclease-mediated genome editing approaches have been reported to last only several months, in some cases even 4 weeks (Yang, Wang, and Jaenisch 2014), significantly reducing the time it takes to generate the transgenic cell or organism.

This is quite a new and exciting technology, for which serious toolboxes have been generated for effective applications and for which design turns out to be a crucial step (Cheng and Alper 2014; Valton et al. 2014).

> Science never solves a problem without creating ten more.
>
> **George Bernard Shaw**

8.8 Problem Session

Answers are available for the questions with an asterisk—see Appendix D.

Q1. You wish to study how the MyoD transcription factor regulates differentiation of the muscle lineage. Would you carry out this analysis using a proliferating myoblast cell line, primary myoblast cell, or primary adult skeletal muscle, and why? If you were to genetically engineer a proliferating myoblast and an adult skeletal muscle in parallel sets of experiments, which vectors would you choose for these studies, and why?

*__Q2.__ You are working in a bioengineering company where your boss asks you to generate a product that mimics the 3D structure of the liver (using embryonic stem cell lines also sold by your company). What would you do if you wanted to *label* the differentiated liver cells with the GFP reporter? Explain.

Q3. You want to modify primary fibroblasts from patients with muscular dystrophy, which is a severe form of muscle degeneration (caused by a mutation in the dystrophin gene, rendering it nonfunctional), so that they stably express a wild-type form of the dystrophin gene. You only have the standard vectors provided in your lab (this book), as well as a cDNA mix from a healthy individual's skeletal muscle. The dystrophin gene sequence is given below. How would you design this direct reprogramming strategy?

5′-ATG GCT ATC GAC TTT CGT ATA GAA ACC AGA CGG GCC GTT CTA GGA TTT GGG TGA

Chapter 9

Genetic Manipulation of Plants

> It might seem unfair to reward a person for having so much pleasure over the years, asking the maize plant to solve specific problems and then watching its responses.
>
> **Barbara McClintock**

In 2008, over 10 million farmers in 25 countries planted transgenic plants, the planted areas increasing from around 44 million hectares in 2000, to 125 million hectares in 2008 (Marshall 2009). The majority of these crops are herbicide-resistant (58.6%), the rest are either insect-resistant or have stacked traits. Although there is still much debate on the ethics of environmental safety, economics, and gene diversity-related issues, transgenic crops are still considered by many as a source of oral vaccines, biofuels, or improved/high-quality food products; hence genetic engineering is either used to create new products in plants or else to assign to plants novel functions, for example, to improve crop quality or quantity.

There are quite a number of ethical issues surrounding genetically modified crop plants (a subset of genetically modified organisms, or GMOs). The main target has long been one of the pioneering companies, Monsanto (St. Louis, Missouri), who is famous for **Bt cotton**, which is a trademark genetically modified cotton that produces an insecticide through expression of a bacterial toxin-producing *cry* gene inside the cotton. *Bacillus thuringiensis* (Bt) *cry* genes code for Cry (crystal) proteins, which are endotoxins that become activated in the acidic pH of the insect stomach, resulting in the death of the insect. Monsanto had, in fact, initially produced the Bt potato and obtained approval from the Environmental Protection Agency (EPA) in 1995, and later on went on to insert the same genes into other plants, producing Bt soybean, Bt corn, and the like. These Bt plants have initially significantly reduced the use of chemical insecticides in the field, resulting in big hype for the biotech industry, although in spite of these insect-resistant GM crops, pesticide use has been estimated to be over 1 billion pounds per year in the United States alone (Alavanja 2009). However, an example in India of Bt cotton indicates no significant yield improvement after Bt cotton implantation, and no significant decrease in pesticide use in Bt cotton-planted areas.* This may or may not be due to misinformation and mishandling of farmers, or due to the Bt cotton itself. In fact, global pesticide sales were steadily increasing all over the world, except perhaps for North America, where Bt cotton has significantly reduced the use of pesticides. However, the amount of pesticide used per hectare in China, for instance, is almost five times more than that used in the United States (Plumer 2013). The reader can explore the ethical issues as well as the financial and economic relations for him- or herself and simply concentrate on the techniques behind how transgenic plants are generated.

* See the Coalition for a GM-Free India document at http://www.biosafety-info.net/file_dir/551137394f82a8adac3ad.pdf.

9.1 Monocotyledons, Dicotyledons, and Commercial Crops

Flowering plants are commonly divided into two classes: monocots and dicots. Although there are some species where a classification is not terribly clear, many model organisms used in plant laboratories can be readily distinguished by a number of features (Table 9.1).

In one sense, monocots can be considered as industrially the most important crops in the world, since the nutrition of the world largely depends on these staple crops. Not surprisingly, much research has been devoted to the genetic

Table 9.1 Basic Distinguishing Features of Monocots and Dicots, with Common Examples

Monocotyledons (Monocots; Liliopsida)[a]	*Dicotyledones (Dicots; Magnoliopsida)*[b]
One cotyledon in the embryo	Two cotyledons in the embryo
One furrow or pore in the pollen	Three furrows or pores in the pollen
Flower petals in multiples of three	Flower petals in multiples of four or five
Scattered stem vascular bundles	Organized stem vascular bundles
Parallel leaf veins	Reticulated leaf veins
No secondary growth	Secondary growth
Roots from nodes in the stem, called *prop roots*	Root from the lower end, the radicle

[a] Rice (*Oryza sativa*), maize (*Zea mays*), barley (*Hordeum vulgare*), sugarcane (*Saccharum officinarum*), banana (*Musa*).

[b] Tobacco (*Nicotiana tabacum, Nicotiana rustica*), tomato (*Lycopersicon esculentum*), thale cress (*Arabidopsis thaliana*), carrot (*Daucus carota*), potato (*Solanum tuberosum*), cotton (*Gossypium*), canola/rapeseed (*Brassica*).

manipulation of these major monocots: rice, maize, barley, and sugarcane. Yet, a lot of work also goes into genetic manipulation of other important food crops such as tomatoes, potatoes, carrots, cotton, and rapeseed (canola), either for improving food quality and content, or improving culture conditions and disease or herbicide resistance, or as biofactories for production of pharmaceutically or industrially important products. Some of the vectors are more suited for the manipulation of monocots, and some for dicots, however, recent advances have aimed at engineering vectors suitable for transformation of both classes of plants. Many transgenic plants of commercial importance have been on the market for decades. We will take a quick a look at a few examples, before discussing plant cell and tissue culture and plant expression and reporter vectors.

The **Flavr Savr tomato** was the first commercial genetically engineered food product approved for human consumption by the Food and Drug Administration (FDA) in 1992. Produced by the company Calgene LLC (Davis, California [now a Monsanto company]), this tomato was rendered more resistant to rotting and softening by transgenic insertion of an antisense gene that suppresses expression from a polygalactouronase gene, which blocks breakdown of pectin in the cell wall, hence reducing softening. However, in practice Flavr Savr tomatoes did not stay firm and had to be reaped just like normal wild-type varieties, hence they did not provide any significant advantage, and did not survive very long in the market.

Golden Rice has been genetically engineered to synthesize beta-carotene in rice by a research team of the Swiss Federal Institute of Technology and University of Freiburg researchers, with the hope of improving the nutritional value of this staple crop for a large percentage of the world population that relies on rice for their diet, and who suffer from vitamin A deficiencies (Ye et al. 2000).

The **Roundup Ready Soybean**, also from Monsanto, has been genetically engineered for increased resistance to the trademark herbicide of Monsanto, glyphosate, which interferes with the synthesis of essential amino acids, and hence is detrimental to not only herbs but also crops. This chemical inhibits an enzyme, 5-enolpyruvylshikimate-3-phosphate synthase (EPSPS), which catalyzes a critical step in the synthesis of these essential amino acids. The Roundup Ready variety is genetically engineered to synthesize a variant of this EPSPS enzyme that is not responsive to glyphosate, which protects the commercial crop from the herbicide. The major problem with the Roundup Ready use has been that although early on the system worked well and herbs were effectively removed with relatively less amount of glyphosate, herbs (similar to bacteria and antibiotics) have gained resistance to this herbicide over the years, and the use of herbicide has gradually increased, effectively outweighing the advantages of the genetic modification (i.e., less use of herbicides—not more).

On a different note, a French team published (alongside many media events prior to the article release) a two-year study they had performed on rats, which were fed a standard diet versus a diet including Roundup herbicide and/or Roundup Ready corn, which they claimed resulted in a high tumor incidence over long-term exposure to this herbicide or the GMO crop (Seralini et al. 2012). Obviously, this led to an uproar by activists against GMO products and GMO regulations all over the world (Arjo et al. 2013). However, when the data were carefully analyzed, it became obvious that the results were inconsistent with the interpretations. For one thing, GM-fed male rats that died of tumors (7 of 30; ~23%) were fewer than those in the control group (3 out of 10; ~30%). Furthermore, there was no dose dependence. Rats fed on an 11% GM-containing diet showed a higher mortality rate than those fed on a 22% GM-containing diet (Arjo et al. 2013; Seralini et al. 2012). As such, the article soon received serious scrutiny from researchers all over the world (Arjo et al. 2013; Ollivier 2013; Sanders et al. 2013), and although the authors have provided lengthy answers to many of these criticisms (Seralini et al. 2013), the article was ultimately retracted.

There have been many more attempts at transgenic crops, some of which are commercialized, some still in the research phase. A Japanese team, for example, has successfully and stably repressed the expression of N-methyltransferase enzymes that participate in caffeine synthesis, thereby generating a transgenically decaffeinated plant (Ogita et al. 2003). In this chapter, we will briefly glance at plant cell or tissue culture systems, manipulation techniques, and some of the more common plant expression vectors used in some of these industrially important plants.

9.2 Plant Manipulation Methods

9.2.1 *Plant Cell and Tissue Culture*

Unlike animal cells, plant cells possess a unique property, they are totipotent—meaning any part of the plant can in theory grow vegetatively and generate an entire plant, which makes both cell and tissue culture as well as manipulation of plants relatively easier than mammalian cells. Many different parts of the plant may be cultured—be it cells from the embryo, specialized organs, callus or dispersed cells. A **callus** is a mass of rapidly proliferating cells at the site where a plant is cut or injured, in a way similar to proliferating precursor cells or fibroblasts at the wound sites in animals. Thus, a callus can develop into root, shoot, and stem structures in tissue culture (Dodds and Roberts 1985).

The same asceptic culture conditions as in mammalian cell culture also apply to plant cell culture. Culture media, however, are based on the nutritional requirements of plant cells and follow a few basic rules, although the media should be optimized for each plant cell. In general, inorganic salts including nitrogen, magnesium, and phosphorus, sugar (mostly sucrose of D-glucose), amino acids, nicotinic acid, glycine and pyridoxine, plant growth hormones (also known as plant hormone regulators, PHRs) such as auxins and cytokinins, and vitamins such as vitamin C, biotin, and riboflavin are added to the culture medium (Dodds and Roberts 1985). Culture medium can either be aqueous or matrix-based (including the starch of sucrose polymers); monocots and dicots may have different culture requirements and detailed protocols should be obtained and optimized in the laboratory.

Plant cells can be genetically manipulated with polycationic chemicals, such as calcium, or liposomes, or electroporation, and so forth. However, the plant cell wall, with its rigid architecture, is one challenge in the genetic manipulation of plant cells, significantly reducing the efficiency of many of the above-mentioned methods of gene delivery. Therefore, several methods have been devised to penetrate the cell wall, thereby increasing the efficiency of transformation.

9.2.2 The Gene Gun

A gene gun, also known as a biolistic particle delivery system or particle bombardment method, is a direct method of nucleic acid delivery into cells that are otherwise difficult to transfer, such as bacteria, yeast, or plant cells. Usually helium-based or another source accelerates gold or tungsten particles coated with nucleic acids, which can then penetrate both the cell membrane and cell wall in such organisms. Once inside the cell, the nucleic acid is removed from the particle and either a transient expression or stable expression (if integrated into the host chromosome, with a very low probability) will take place.

9.2.3 Protoplasts

The highly structured and rigid cell wall of plants is a serious obstacle for the delivery of DNA; therefore, in some strategies the cell is stripped of this cell wall, and is left with only its plasma membrane. This is called the **protoplast**. The first protoplasts were cultured in 1892 from onion bulb scales by Klercker; however, it was not until the 1960s that the method was refined for sterile tissue culture applications (Compton et al. 1996). A protoplast can be obtained using two major protocols—(a) either plant tissues can be mechanically sliced or chopped, which would damage many of the cell walls (the original method which will also result in the loss of many cells, hence it is not very popular), or (b) a milder version that uses hydrolytic enzymes to get rid of the cell wall, while maintaining the cell membrane (the preferred and more common version). As with other plant cultures, ammonium nitrate and calcium is often needed to promote cell division, along with organic components, sugar, plant growth hormones, and so forth, however, the culture conditions need to be optimized for protoplasts. Protoplasts are usually sensitive to light; hence they should be maintained in dark conditions until they synthesize a new cell wall (Compton et al. 1996).

A protoplast, devoid of its cell wall and left with only a cell membrane, can then be transfected with DNA the same way as a typical mammalian cell—for example, by electroporation or polycationic chemicals, liposomes, and so forth. Protoplasts can also be co-cultivated with *Agrobacterium tomafeciens* (Compton et al. 1996).

9.2.4 Agrobacterium

A. tumfaciens is a plant pathogen that causes **crown gall disease**, a type of plant tumor. These bacteria transfer a tumor-inducing (Ti) plasmid to a wide range of monocot or dicot

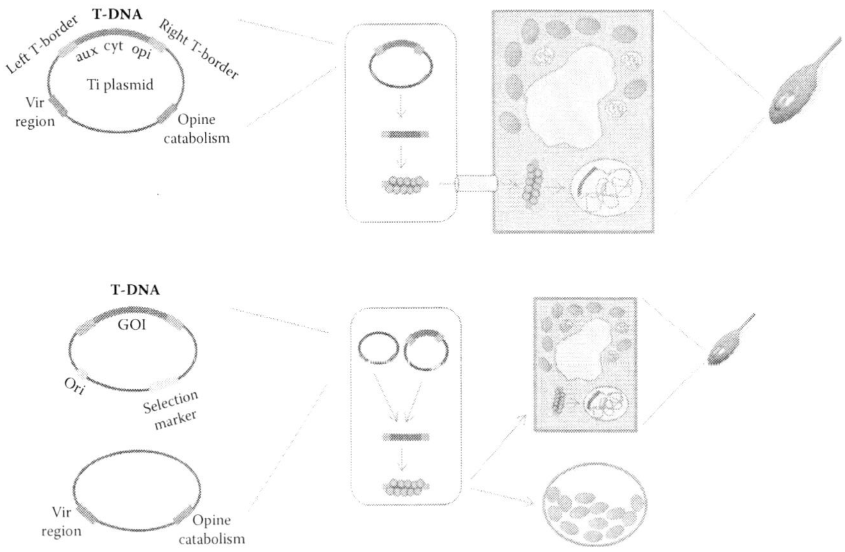

Figure 9.1 A simple schematic of an *Agrobacterium*-mediated plant transformation. *Agrobacterium* typically transfers its T-DNA region (coated with proteins encoded by the *vir* region) to the wounded leaf, causing overproliferation, hence, *tumor-induction* (upper panel). This feature is exploited in plant genetic manipulation: a disarmed Ti plasmid, which contains a virulence region but no T-DNA (hence, it cannot *infect* the plant) and a binary vector consisting of the T-DNA left and right borders with the gene of interest (GOI) cloned in between, a bacterial replication origin (ori), and a bacterial selection marker (for easy cloning in *E. coli)* are co-transformed into the same *Agrobacterium*, which then uses the proteins coded by the *vir* region *in trans* to coat the transgene flanked by the left and right borders— this mechanism can be used either on wounded plants as shown, or on protoplast cultures (lower panel). See the text for details.

plants. This Ti plasmid contains a so-called T-DNA (transferred DNA), with genes for tumor induction and nopaline synthesis (Figure 9.1). These T regions on the Ti plasmid, defined by left and right T-DNA borders, are usually 10 to 30 kbp in size, and there may be multiple T regions on some Ti plasmids (Gelvin 2003). The virulence endonucleases recognize and cut these border sequences, releasing the T region, and the single-stranded T strand is then coated with the VirD2 protein and gets transferred to the plant with the help of a combination of other virulence proteins (Figure 9.1).

For *Agrobacterium*-mediated transformation, protoplasts are typically isolated for a few days before the day of transfer, and then co-cultivated with the genetically modified *Agrobacterium* (usually at a 1 plant cell:100 bacteria ratio). After co-cultivation antibiotics are added to the culture media so as to eliminate bacteria, and a selection marker is used to select for transformed plant cells (Gelvin 2003).

Historically speaking, cointegrate Ti plasmids (or hybrid Ti plasmids) were the first vector systems used, however, they are not widely employed today, as easier-to-manipulate systems have been devised. More commonly, engineered T-DNA binary vectors are used in *Agrobacterium*-mediated transformation, which contains two separate vectors—one that contains a helper Ti plasmid that lacks the T-DNA but contains virulence genes, and a binary vector where the T-DNA is present on a smaller binary vector, where the transgene is cloned between the left and right borders, but there is no virulence region (sometimes also referred to as the *mini-Ti*). This mini-Ti can also be cloned in *Eschericia coli* for easier manipulation, and then transformed into an *Agrobacterium* that harbors the disarmed helper Ti plasmid that provides the *vir* genes *in trans*.

9.2.5 Plant Expression and Reporter Vectors

The method of DNA delivery for some cases also defines the types of expression plasmids that one can choose (Liu et al. 2013). The common point in all of these plant expression plasmids, however, is the same as before:

a. A **promoter** that is strongly active in plant cells. Many of the plant expression plasmids use promoters from plant viruses, such as those of the cauliflower mosaic virus (CaMV) genes.

One may also use tissue-specific promoters as before, such as a fruit-specific, leaf-specific, or seed-specific promoter, depending on where the transgene expression is required.

b. **Selectable markers** such as *nptII* (encoding resistance to kanamycin), *hptII* (encoding resistance to hygromycin) or *pat* (encoding phosphinothricin N-acetyltransferase, which detoxifies phosphinothricin (ppt), an inhibitor of plant growth). In many of the food crops, selectable markers are removed (mostly through Cre-LoxP, Flp-FRT, or similar site-specific recombinase-mediated cleavage of the marker gene and/or the transgene) prior to market release due to biosafety concerns, but this also allows the producer to not label its product as "GMO."

c. **Reporter genes** such as *gusA* (β-glucuronidase, which will cleave the X-gluc, 5-bromo-4-chloro-3-indolyl glucuronide, which is a colorless substrate, into 5,5′-dibromo-4,4′-dichloro-indigo, a blue-colored insoluble product), *smgfp* (a soluble version of the codon-modified green fluorescent protein, the so-called soluble-modified GFP) (Davis and Vierstra 1998), and *luc* (encoding the enzyme luciferase).

pCAMBIA vectors are among the most commonly used plant expression vectors due to the high-copy number (hence the high DNA yield in *E. coli*), high stability in *Agrobacterium*, ease of cloning the gene of interest, and ease of selection in both bacteria and plants (Figure 9.2).

The GATEWAY system for plant expression uses the classical GATEWAY technology (Chapter 2, Section 2.3) for effective and easy cloning of large DNA sequences into T-DNA binary vectors, which can otherwise be time consuming (Karimi, Inze, Depicker 2002). Various versions of GATEWAY system vectors are possible for overexpression, GFP fusion synthesis, gene silencing, marker expression, promoter analysis, and so forth, in plants (Karimi, Inze, Depicker 2002). There is also a new series of modular binary vectors that could be used with

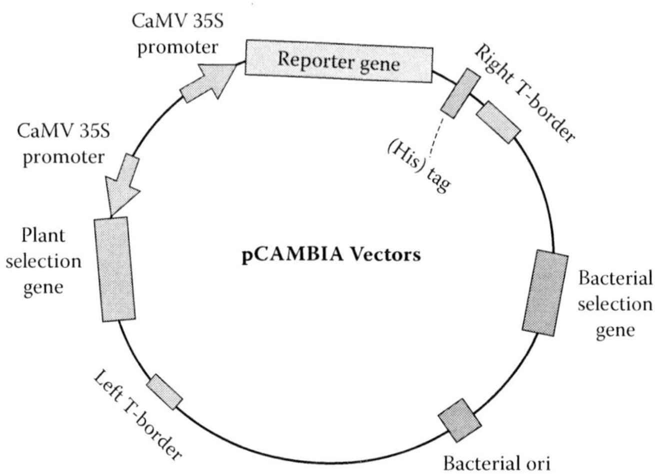

Figure 9.2 A diagram of a generic pCAMBIA vector, which expresses a reporter gene and a plant selection gene from their respective CaMV 35S promoters, flanked by left and right T-borders. The His tag helps affinity purification of the expressed protein, if required. Bacterial ori and bacterial selection genes are used for cloning in bacteria.

both *Agrobacterium*-mediated transformation and gene gun strategies in monocots as well as dicots, such as the pORE (Coutu et al. 2007).

Plant virus-based gene silencing vectors are also available, some of which are based on plant viruses such as the bamboo mosaic virus (BaMV) or tobacco rattle virus (TRV) (Liou et al. 2014). Virus-induced gene silencing (VIGS) is quite common in downregulating expressions from genes in plants, exploiting the natural antiviral defense mechanism (Purkayashta and Dasgupta 2009).

It should be remembered that plant cells have three genomes: nuclear, mitochondrial, and plastid (mainly chloroplast). Manipulation of the chloroplast genome has been realized in a limited number of plant species, starting with tobacco, and has attracted some attention due to its many advantages, including exclusively homologous recombination-mediated integration, a high number of chloroplasts (hence the high levels of stable expression in green tissues, although there are problems with expression in nongreen tissues such as roots or seeds), lack of gene silencing mechanisms, and maternal mode of inheritance that significantly reduces transgene transmission through pollen (Bock 2014). The gene gun is the common method used, since the transgene has to penetrate several layers of membrane, however, plastid transformation methods are still not well established (Bock 2014).

Another point to remember is that codon usage may change from bacteria to fungi to humans to plants, hence for optimum translational efficiency of the transgene in plants, codon optimization is recommended. If, for example, a human gene is to be expressed in plants, one must analyze the codon frequencies coding for the same amino acid in humans and plants. Supposing the protein product that has the amino acid glycine is coded by GGU in humans, with a codon frequency of 14.5%, but in plants this codon has a frequency of 2%. This would interfere with the translational efficiency. However, the codon GGC, again coding for glycine, has a frequency of 30% in plants. Then, replacing GGU with GGC would improve translational efficiency significantly in plants. This is particularly true for organelle genomes, where codon frequencies tend to show greater variation than nuclear genomes.

Needless to say, the plant genetic manipulation methods discussed here only cover the tip of the iceberg in the field. However, we hope that the chapter provides the reader with a rough idea about the plethora of things that can be done in plants, from biopharmaceuticals to biofuels.

9.3 Future Trends in Transgenic Plants

By 2010, almost 150 million hectares of the world were planted with genetically modified crop plants according to a report by the Food and Agriculture Organization (FAO at http://www.fao.org/docrep/015/i2490e/i2490e04d.pdf). These not only included insect- or herbicide-resistant cotton, canola, maize, and so on, or nutritionally fortified crops such as the Golden Rice discussed previously, but also included transgenic plants engineered for phytoremediation purposes, or as bioreactors for pharmaceuticals, such as bananas producing vaccines. In general, the United States, Canada, and Argentina appear to be the major transgenic or GM crop producers and exporters; and in fact although not commercialized at a large scale, the global value for GM rice is estimated to be around 64 million USD per year, which implies that it is an important staple crop (Demont and Stein 2013). Biofuel and bioenergy sectors are the next big sectors in transgenic research, where the United States and Brazil are the largest producers (FAO at http://www.fao.org/docrep/015/i2490e/i2490e04d.pdf).

Almost 25 years after the first genetically modified food crop, Flavr Savr, was produced gene targeting in plants is still being updated, improved, and field-tried—zinc finger nucleases (ZFNs) and other new genome editing tools have made gene targeting significantly easier not only in animals, but also in plants (Puchta and Fauser 2013), which (when combined with synthetic biology approaches that will be discussed in Chapter 10) may open up new avenues for transgenic research. The green revolution is far from over—it may only be just beginning.

9.4 Problem Session

Answers are available for the questions with an asterisk—see Appendix D.

***Q1.** You are an agricultural researcher who is trying to use rice as bioreactors to express the industrial enzyme XYZ from tobacco plants (the gene sequence is shown below):

5′- ATG CGC GCC ATT ATA TCT GGT CGA ACC TGC ACC TAA –3′

How would you design the experiment to produce the transgenic rice producing this enzyme, for later purification process in your company? Design the R&D process, confirming that you indeed express the enzyme.

Q2. Omega-3 long-chain polyunsaturated fatty acids (ϖ3, mainly DHA) have critical roles in human health and development with studies indicating that deficiencies in these fatty acids can increase the risk or severity of cardiovascular and inflammatory diseases in particular. Your company wishes to produce this fish oil in *Arabidopsis thaliana* seeds for commercial use.

Q3. The critical pathways in the production of fish oil (DHA) are shown on the left, the genes coding for the two main enzymes Δ5 elongase (Δ5 elo) and Δ4 desaturase (Δ4des), as well as the seed-specific promoter FAE1, are given on the right:

Eicosapentaenoic acid, EPA Δ5-elo ↓ $22{:}5^{\Delta 7,10,13,16,19}$ Docosapentaenoic acid, DPA Δ4-des ↓ $22{:}6^{\Delta 4,7,10,13,16,19}$ Docosahexaenoic acid, DHA	Δ 5 elo : 5′- ATG CGC GCC ATT ATA TCT GGT CGA ACC TGC ACC TAA –3′ Δ 4 des: 5′- ATG GGG CTT AAA ATA CTA GAG GCC GAT TTT ATA TAG –3′ FAE1 promoter: 5′- atcgtagctagcatcgtagtagc atcgtagcgtagctagctgatcg- 3′

Design a strategy to engineer such an improved *A. thaliana* plant.

Q4. You know that the chloroplast TaXH gene improves salt tolerance in plants such as potatoes or peppers; you wish to use this gene to engineer a salt-tolerant cranberry plant. The potato chloroplast TaXH (pcTaXH) gene sequence is given below. Design a strategy to generate salt-resistant cranberries.

PcTaXH 5′-ATG CAG GGG AGA TTT TCC......... CAT CGA ACA GAC CGA CGA TGA –3′

Q5. You wish to increase the provitamin A content in plant plastids, so as to improve food quality. In order to do this, you need to engineer the carotenoid metabolic pathway, by overexpressing lycopene β-cyclase gene from daffodil in your target plant, rice. However, you only have the plastid reporter vector on the right (which contains no Ti-derived element; you got it from the author of the paper) for expression of the reporter gene in chloroplast. How would you design this metabolic engineering using the vector at hand?

Lycopene β-cyclase gene sequence:

5′- ATG CAA ATA TTT CCC CGG GAA GCA............GTT TCC CGT ATC AGT CTA TAG –3′

Chapter 10

Today and the Future

The science of today is the technology of tomorrow.

Edward Teller

Molecular biology as a field was born in the late 1930s from the marriage of physics and chemistry, in an effort to explain nature, and gained momentum during the postwar atmosphere of the 1940s. With the advent of nuclear physics, radioactivity, quantum mechanics, and X-ray crystallography, it became easier to study the actions and mechanisms of biological macromolecules, and after the structure of DNA was discovered in 1953 (Watson and Crick 1953) (also see Chapter 1 for a brief history), efforts in molecular biology accelerated even more.

Also during the postwar era, another field was born out of improved computation efforts due to the extensive race in cryptography during the war. Enigma, developed by German engineers toward the end of World War I, was used to encrypt German military texts. Alan Turing was a cryptanalyst who worked for the British code-breaking center during World War II, and is, in fact, the father of today's computers. His Turing machine, invented by Turing in 1936, is considered to be the prototype computer. The first freely programmable computer, the Z1 computer, was developed by Konrad

Zuse (the Z3 model was finished in 1941), followed by the ABC computer built by Atanasoff and Berry in 1942, and the famous ENIAC computer (Electronic Numerical Integrator And Computer) developed by the MIT team in 1943 (see the Timeline of Computer History at http://www.computerhistory.org/timeline/?category=cmptr). The invention of the transistor in 1947 and the integrated circuit in 1958 greatly affected computer technology. The Advanced Research Projects Agency Network (ARPANET), which in a sense was the early Internet prototype that linked computers at Stanford, UCSB, UCLA, and the University of Utah, was established in 1969.

So in a sense it is possible to say that the two sister fields, born in the same postwar era, developed hand-in-hand and supported each other.

10.1 Bioinformatics and the Omics Age

When Frederick Sanger developed his DNA sequencing technique (see Appendix A), the infrastructure for a genome sequencing frenzy was all set. The Institute for Genetic Research (TIGR) established the first genomic sequence of an organism, *Haemophilus influenza*, with the help of the new computational tools developed at this institute. In 1989, a European consortium was set up to sequence the yeast genome, and in 1990 the Human Genome Project was announced. The entrepreneur and scientist, J.C. Venter, should be mentioned here, since his team at TIGR, as well as his later-founded company Celera Genomics, developed the shotgun sequencing technology, which greatly improved the speed of these genome sequencing efforts.

The problem with genome sequencing projects was twofold—the technology of DNA sequencing (especially before shotgun sequencing and automated sequencing approaches) was extremely slow, so that an international effort involving thousands of labs and researchers was required to even attempt the sequencing of the human genome. The second problem, however, was more of not only acquisition of data, but also of storage, analysis, and manipulation of data, as well as access from all over the world.

The Protein Data Bank (PDB) was already founded in 1972, to store X-ray crystallography data of proteins, whereas it took over 10 years to establish the Swiss-Prot protein sequence database. The National Center for Biotechnology Information (NCBI), as well as the Human Genome Organization (HUGO), was founded in 1988. The corresponding European effort for a nucleotide sequence database, the so-called European Bioinformatics Institute (EBI), had its roots in the European Molecular Biology Laboratory (EMBL) Nucleotide Sequence Data Library, established in 1980.

The genome sequencing approaches, now termed **genomics**, soon expanded exponentially, with model organisms sequenced one by one, from *Saccharomyces* to *Arabidopsis*, from rice to mouse. Analysis of this data was also a huge task, which included sequence alignments, comparisons, analysis, generating tens and hundreds of different online tools, and also creating new and more specialized avenues such as comparative genomics, functional genomics, structural genomics, pharmacogenomics, metagenomics, and so on and so forth. Soon thereafter, with the advance in microarray technologies that made global gene expression profiling possible, whole *transcriptomes* became popular. Genes were important, yes, but genes were not always expressed, or expressed to the same level, in every cell type at all times. Therefore, **transcriptomics** was the next "it" thing in molecular biology. Very soon it became clear that just because a gene is transcribed, it does not mean that the protein product is produced—mRNAs could become degraded before being translated, or translation could be stalled by other means. Therefore, *whole proteome profiling*, otherwise known as **proteomics**, became popular. All these *omics* also meant an exponential increase in the amount of data generated and this added to the quantity of bioinformatics data, wherein new analysis tools became necessary. That was the circle of research life. Now, other omics are also in fashion, creating an ever-growing mountain of data, from **lipidomics** to **metabolomics**, all of which should somehow be integrated so as to give useful and informative knowledge of the molecular mechanisms of cells.

The Biology Workbench analysis environment, maintained at the San Diego Supercomputer Center (SDSC), is an all-in-one and easy-to-use platform for nucleotide and protein sequence data analysis, including restriction enzyme analysis, protein structure predictions, *in silico* translations, sequence alignments, and much more, with access to other well-known databases such as NCBI or Swiss-Prot from within the platform (Subramaniam 1998) (Biology Workbench at http://workbench.sdsc.edu). The access is free, only a registration is required for the maintenance of data; however, unfortunately due to an apparent lack of funding it runs very slowly and new and improved versions are not currently being released. Biology Workbench remains a comprehensive analysis tool for the novice student, with quite an extensive coverage of a multitude of sequence analyses. Nevertheless, new analysis tools are constantly being developed, especially with the accumulation of the wealth of data from high-throughput studies, with more user-friendly and visualized interfaces (Gehlenborg et al. 2010). Such tools mostly analyze microarrays, transcriptomics, and proteomics data and generate protein interaction networks, pathway analyses, and other approaches so as to generate biologically relevant information. **Systems biology** is an interdisciplinary, or rather a *cross-disciplinary*, field that combines information obtained from all other omics approaches so as to present an integrated and interacting network of molecules, genes, proteins, and pathways, studying the *system* as a whole, rather than focusing on a number of molecules through a keyhole, as well as on system dynamics. The Institute for Systems Biology (ISB at http://www.systemsbiology.org) was founded in 2000, with exactly this in mind, and almost a decade and a half later this institute has made its impact with research projects on the environment, brain, and personalized medicine, among many others. The Web portal for systems biology researchers is also freely available at http://systems-biology.org, funded by the Systems Biology Institute of Japan, also founded in 2000 (http://sbi.jp).

> There are living systems; there is no living "matter." No substance, no single molecule, extracted and isolated from a living being possess, of its own, the aforementioned paradoxical properties. They are present in living systems only; that is to say, nowhere below the level of the cell.
>
> **Jacques Monod**

10.2 Synthetic Biology and Unnatural Amino Acids

Synthetic biology is defined as:

(A) The design and construction of new biological parts, devices, and systems, and
(B) The re-design of existing, natural biological systems for useful purposes,

in the synthetic biology community portal (http://syntheticbiology.org). Synthetic biology is a recently new interdisciplinary and transdisciplinary field, combining biology, physics, bioinformatics, engineering and many other fields, in order to use biological systems as, in a manner of speaking, a "chassis" and molecules as "Lego pieces," rearranging these pieces to come up with new uses that will (hopefully) benefit society. The International Genetically Engineered

Machine (iGEM at http://igem.org) competition was initiated by MIT researchers in 2003 and has since been the "in" competition, where new biological system designs compete every year (especially noteworthy is the iGEM high school jamboree). The same MIT team also introduced the BioBrick, which is a standardized platform of parts, devices, and systems for easy assembly of synthetic systems (originally the Registry of Standard Biological Parts at http://parts.igem.org/Main_Page), which is freely available (for instance, BioBrick® Assembly Kit at http://www.neb.com/products/e0546-biobrick-assembly-kit). It should be noted that the MIT Center for Integrative Synthetic Biology has also developed the CRISPR/Cas system for genome editing, as discussed previously.

Synthetic metabolic or genetic circuits, new protein designs, synthetic biosensor systems, and many other applications have been engineered through synthetic biology. A recombinase-based platform for logic and memory functions was engineered using such a synthetic biology approach, for long-term and stable maintenance of cell memory by an integrated circuit of AND and OR gates (Siuti et al. 2014). A synthetic biology toolbox for yeast as a biofactory has also been developed (Redden, Morse, and Alper 2014), which could be expected to help design new synthetic circuits for the production of high-value yeast metabolites which are otherwise biochemically difficult to obtain in large quantities (Dai et al. 2014). Synthetic biology is also available for engineering higher eukaryotic systems, such as plants and mammals, for the production of commercially important metabolites or therapeutic proteins (Wilson, Cummings, and Roberts 2014; Ye and Fussenegger 2014).

Perhaps one of the most exciting developments in synthetic biology, however, was the *expansion of the genetic code with unnatural amino acids* (where the author admits to being personally a little subjective here). Schultz Laboratory has been pioneering in the field, as they have engineered bacteria, yeasts, and mammalian neurons that harbor an expanded genetic code. The Standard Genetic Code contains 61 codons (leaving the three stop codons) that code for 20 amino acids—however, certain organisms, such as bacteria, archaea, or yeast, may sometimes use one of the stop codons to code for a 21st (or sometimes a 22nd) "unnatural" amino acid, such as a selenocysteine (Sec) or a pyrrolysine (Pyl), using either a SECIS (Sec insertion sequence) or PYLIS (Pyl insertion sequence) elements (Yuan et al. 2010; Zhang et al. 2005). (In humans, only 25 selenoproteins are found.) This has been the starting point of the Schultz lab, and they have redesigned the bacterial translation machinery (initially) in such a way that one of the stop codons is reprogrammed to incorporate other unnatural amino acids (Wang and Schultz 2002). The Schultz group thereafter carried out or collaborated on a number of different synthetic biology projects involving unnatural amino acids, such as protein evolution, where they have shown synthetic codes can confer selective advantage to antibody proteins (Liu et al. 2008). They have also analyzed a collection of 138 unnatural amino acids, and a majority of glutamine and glutamic acid analogs were determined to have uptake by yeast, giving way to expansion of the genetic code in this eukaryotic model organism (Liu and Schultz 1999). Five unnatural amino acids were successfully incorporated to the yeast TAG codon, whereby the keto group on these unnatural amino acids make photocross-linking for protein interaction studies possible (useful in photoclick chemistry) (Chin et al. 2003). And, synthetic engineering of the unnatural amino acid p-acetylphenylalanine (pAcF) was incorporated into the human growth hormone (hGH), with comparable clinical performance to wild-type (Cho et al. 2011).

However, this was not sufficient for the Schultz Laboratory, and they once again expanded the horizon when they engineered *Eschericia coli* with a functional quadruplet codon (AGGA) coding for L-homoglutamine (hGln) (Anderson et al. 2004). This by itself implied one could now imagine previously unimaginable new synthetic genomes and cells, with new and improved functions, and it was a huge breakthrough. Following on that note, recently a different team has engineered a "semisynthetic" *E. coli* that contain an **unnatural genetic alphabet**, incorporating unnatural nucleotide triphosphates d5SICS and dNaM to the genome, which base pair, efficiently PCR-amplify, get transcribed, and are not removed by DNA repair machinery (Malyshev et al. 2014). One can be pretty certain that such unnatural nucleotide triphosphates will soon be incorporated to eukaryotic genomes, greatly expanding the genetic manipulation toolkit.

> Your theory is crazy, but it's not crazy enough to be true.
>
> **Niels Bohr**

10.3 Optogenetics

Synthetic biology has not only expanded the amino acid and nucleotide toolkit, but has also expanded genetic circuit design horizons, generating many blue-, red-, UV-, and other light-controlled circuits (Bacchus et al. 2013), which has made the scientific atmosphere suitable for the birth and maturation of the optogenetics field.

Optogenetics, developed originally in the Deisseroth Laboratory, was chosen as the Method of the Year 2010 (Deisseroth 2011; Editorial 2011). The official resource Web page from the Karl Deisseroth Laboratory is available from http://web.stanford.edu/group/dlab/optogenetics.

This technology is in fact a beautiful and elegant example of how big ideas actually do not require so much an exceptional genius per se, but the capacity to see the previously unseen. The idea that light could be used to precisely control neural activity in specific cell types of the brain was originally proposed by Francis Crick in the late 1970s; however, it was not until 2002 onward that rhodopsins were used to genetically target specific neurons in *Drosophila* to generate light-sensitive responses (Zemelman et al. 2002; Zemelman et al. 2003). This was the same group that reported the first light-induced control of wing beating, flight, and so forth, behavior in *Drosophila*, whose neurons have been genetically modified to express photosensitive channels (Lima and Miesenbock 2005). The system was then quickly adapted to other model organisms such as *Caenorhabditis elegans* and mice.

Optogenetics is principally based on light-sensitive proteins of microorganisms, mainly channelrhodopsin, halorhodopsin, and archaerhodopsin. Bacteriorhodopsins have been known since the 1970s; halorhodopsin was discovered in the late 1970s. Yet, it was the discovery of channelrhodopsin in the early 2000s that truly opened the gate for optogenetics (Deisseroth 2011). These light-sensitive channels can be turned on or off in response to different wavelengths, upon which channels either open or close (depending on which engineered variety is used), resulting in electrical excitation or inhibition of the neuron in question.

When these light-sensitive channels are cloned downstream of a neural cell type-specific promoter and genetically knocked in to the organism (see commonly used viral vectors, as described in Chapter 8), the only other components of optogenetics required are the optical stimulation (through integrated fiberoptic and solid-state light sources for application in freely moving animals, usually directly mounted to the animal's skull) and usually another genetically knocked-in reporter (biosensor) that can provide rapid readouts, such as the voltage-sensitive fluorescent protein (VSFP) (Deisseroth 2011; Gautier et al. 2014; Knopfel et al. 2010).

Optogenetic reporters have also been adapted for a more general reporter function in genetic engineering, with a broad range of optogenetic reporters to study live-cell imaging of signaling and metabolism; while the first generation optogenetic reporters included calcium reporters and neural activators, second generation reporters have come to include voltage reporters, as discussed previously, neural silencers, and many others (Alford et al. 2013; Knopfel et al. 2010). Tetracycline-inducible systems have also been adapted to optogenetics, increasing the level of control over which neurons are activated and when (Tanaka et al. 2012). Optogenetics has thus gained immense popularity in behavioral and molecular brain research, especially in the areas of learning and memory (rev. in Gautier et al. 2014 and Goshen 2014). Optogenetics approaches have been studied for retinal prosthesis for patients that require restoration of vision (Barrett et al. 2014). They have also been adapted to primates, implying human clinical applications in the near future (Gerits and Vanduffel 2013).

Optogenetics is not only limited to basic neuroscience, although it finds quite a large niche in this field. Cardiac optogenetics aims at optical control of bioelectricity and effective treatment of rhythm disorders (Boyle, Entcheva, and Trayanova 2014). Similar applications are also possible in glial cells and skeletal muscle cells, all of the above having the common feature of electrical excitability, making the use of light-sensitive ion channels suitable for precise and cell-specific regulation of function, although electrically nonexcitable cells are also under investigation as potential new optogenetic targets (Tanaka et al. 2012). In addition, integration of optogenetics with unnatural amino acid incorporation and phenyl azide chemistry makes click chemistry-based novel post-translational modifications possible (Reddington et al. 2013), and one might expect to hear more from such integrated approaches in the future.

10.4 What Is Next?

It is difficult to foresee what the future holds in genetic engineering. However, it has become increasingly clear that new technologies will be developed at a much faster rate than one can process the data generated by those technologies, and that research is bound only by man's imagination.

Nonetheless, it is evident that scientists are still inspired by nature's myriad of miracles, best exemplified by optogenetics. The recent connectomics technology in neuroscience appears to be the next big chapter in things to come. So in conclusion, undergraduate readers of this book should not only focus on their immediate projects at hand, but also continue reading and searching for unknown territories, studying perhaps previously unstudied or neglected avenues, and find a way to incorporate them into today's technologies. What is next? ... Who knows?

> Everything is theoretically impossible, until it is done.
>
> **Robert A. Heinlein**

10.5 Problem Session

Q1. Optogenetics is a recently emerged field whereby one can genetically modify specific neurons to function in response to specific light stimuli. The optogenetic tools are based on microbial opsins that respond to brief pulses of light. Your company is trying to generate a transgenic rat that expresses light-sensitive channelrhodopsin (CatCh, already cloned in the pCMV-HA vector in the figure for storage) in *dopaminergic neurons* for commercial use in neuromarketing research. Also given in the figure is the sequence of the dopamine-specific TH gene. Design a strategy for generating such a transgenic line.

EcoRI **Sal I**

GAG GCC CG AATT CGG GAA AGA TTT CTT CAA............CatCH coding sequence............CGG GTT GAG CTT ATA TG A ATT CGG TCG ACC

−900 +1

............ATCGATCGATCGTACGGGCATCGATCGATCGA............TH promoter sequence............CGATCGTAGCTAGCTGATTATATTCACGAGTATATACA......

References

Alavanja, M.C.R. (2009). Pesticides use and exposure extensive worldwide. *Rev Environ Health* 24(4): 303–309.

Alford, S.C., Wu, J., Zhao, Y., Campbell, R.E., and Knopfel, T. (2013). Optogenetic reporters. *Biol Cell* 105(1): 14–29.

Anderson, J.C., Wu, N., Santoro, S.W., Lakshman, V., King, D.S., and Schultz, P.G. (2004). An expanded genetic code with a functional quadruplet codon. *Proc Natl Acad Sci* 101(20): 7566–7571.

Arjo, G., Portero, M., Pinol, C., Vinas, J., Matias-Guiu, X., Capell, T., Bartholomaeus, A., Parrott, W., and Christou, P. (2013). Plurality of opinion, scientific discourse and pseudoscience: An in-depth analysis of the Seralini et al. study claiming that Roundup Ready corn or the herbicide Roundup cause cancer in rats. *Transgenic Res* 22: 255–267.

Bacchus, W., Aubel, D., and Fussenegger, M. (2013). Biomedically relevant circuit-design strategies in mammalian synthetic biology. *Mol Syst Biol* 9: 691.

Barrett, J.M., Berlinquer-Palmini, R., and Degenaar, P. (2014). Optogenetic approaches to retinal prosthesis. *Vis Neurosci* 6: 1–10.

Beil, J., Fairbairn, L., Pelczar, P., and Buch, T. (2012). Is BAC transgenesis obsolete? State of the art in the era of designer nucleases. *J Biomed Biotech* vol. 2012, article ID 308414, doi: 10.1155/2012/308414.

Berg, J.M., Tymoczko, J.L., and Stryer, L. (2002). *Biochemistry*, 5th Ed. New York: W.H. Freeman.

Bhattacharya, B., Miura, T., Brandenberger, R., Mejido, J., Luo, Y., Yang, A.X., Joshi, B.H., Ginis, I., Thies, R.S., Amit, M., Lyons, I., Condie, B.G., Itskovitz-Eldor, J., Rao, M.S., and Puri, R.K. (2004). Gene expression in human embryonic stem cell lines: Unique molecular signature. *Blood* 103(8): 2956–2964.

Bock, R. (2014). Genetic engineering of the chloroplast: Novel tools and new applications. *Curr Opin in Biotech* 26: 7–13.

Bolivar, F., Rodriguez, R.L., Greene, P.J., Betlach, M.C., Heyneker, H.L., Boyer, H.W., Crosa, J.H., and Falkow, S. (1977). Construction and characterization of new cloning vehicles. II. A multipurpose cloning system. *Gene* 2: 95-113.

Boyle, P.M., Entcheva, E., and Trayanova, N.A. (2014). See the light: Can optogenetics restore healthy heartbeats? And if it can, is it really worth the effort? *Expert Rev Cardiovasc Ther* 12(1): 17–20.

Braam, S.R., Denning, C., van den Brink, S., Kats, P., Hochstenbach, R., Passier, R., and Mummery, C.L. (2008). Improved genetic manipulation of human embryonic stem cells. *Nat Methods* 5: 389–392.

Brophy, B., Smolenski, G., Wheeler, T., Wells, D., L'Huillier, P., and Laible, G. (2003). Cloned transgenic cattle produce milk with higher levels of β-casein and κ-casein. *Nature Biotech* 21: 157–162.

Castro, F.O., Toledo, J.R., Sanchez, O., and Rodriguez, L. (2010). All roads lead to milk: Transgenic and non-transgenic approaches for expression of recombinant proteins in the mammary gland. *Acta Sc Vet* 38(suppl. 2): s615–s626.

Cheng, J.K. and Alper, H.S. (2014). The genome editing toolbox: A spectrum of approaches for targeted modification. *Curr Opin Biotechnol* 30C: 87–94.

Chien, A., Edgar, D.B., and Trela, J.M. (1976). Deoxyribonucleic acid polymerase from the extreme thermophile *Thermus aquaticus*. *J Bact* 174: 1550–1557.

Chin, J.W., Cropp, T.A., Anderson, J.C., Mukherji, M., Zhang, Z., and Schultz, P.G. (2003). An expanded eukaryotic genetic code. *Science* 301: 964–967.

Cho, H., Daniel, T., Buechler, Y.J., Litzinger, D.C., Maio, Z., Putnam, A.M., Kraynov, V.S., Sim, B.C., Bussell, S., Javahishvili, T., Kaphle, S., Viramontes, G., Ong, M., Chu, S., Becky, G.C., Lieu, R., Knudsen, N., Castiglioni, P., Norman, T.C., Axelrod, D.W., Hoffman, A.R., Schultz, P.G., and Kimmel, B.E. (2011). Optimized clinical performance of growth hormone with an expanded code. *Proc Natl Acad Sci* 108(22): 9060–9065.

Cohen, S., Chang, A., and Hsu, L. (1972). Nonchromosomal antibiotic resistance in bacteria: Genetic transformation of *Escherichia coli* by R-Factor DNA. *PNAS* 69(8): 2110–2114.

Collins, J. and Bruning, H.J. (1978). Plasmids useable as gene-cloning vectors in an *in vitro* packaging by coliphage lambda: "Cosmids." *Gene* 4: 85–107.

Colman, A. (1999). Dolly, Polly and other "ollys": Likely impact of cloning technology on biomedical uses of livestock. *Genet Anal: Biomol Eng* 15(5): 167–173.

Compton, M.E., Saunders, J.A., and Veilleux, R.E. (1996). Use of protoplasts for plant improvement (Ch 23), in *Plant Tissue Culture Concepts and Laboratory Exercises* (Trigiano, R.N. and Gray, D.J., Eds.). Boca Raton, FL: CRC Press.

Coutu, C., Brandle, J., Brown, D., Brown, K., Miki, B., Simmonds, J., and Hegedus, D.D. (2007). pORE: A modular binary vector series suited for both monocot and dicot plants. *Transgenic Res* 16(6): 771–781.

Crick, F.H.C., Brenner, S., Barnett, L., and Watts-Tobin, R.J. (1961). General nature of the genetic code for proteins. *Nature* 192: 1227–1232.

Dai, Z., Liu, Y., Guo, J., Huang, L., and Zhang, X. (2014). Yeast syn thetic biology for high-value metabolites. *FEMS Yeast Res*, doi: 10.1111/1567-1364.12187.

Davidson, B.L. and McCray, P.B. (2011). Current prospects for RNA interference-based mechanisms. *Nat Rev Genet* 12: 329–340.

Davis, S.J. and Vierstra, R.D. (1998). Soluble, highly fluorescent variants of green fluorescent protein (GFP) for use in higher plants. *Plant Mol Biol* 36(4): 521–528.

Deisseroth, K. (2011). Optogenetics. *Nature Methods* 8: 26–29.

Delbruck, M. (1949). A physicist looks at biology. *Trans Conn Acad* 38: 190.

Demont, M. and Stein, A.J. (2013). Global value of GM rice: A review of expected agronomic and consumer benefits. *N Biotech* 30(5): 426–436.

Di Berardino, M.A. (2001). Animal cloning—The route to new genomics in agriculture and medicine. *Differentiation* 168: 67.

Docstoc. Notes Amino Acids and Proteins. *Docstoc*, http://www.docstoc.com/docs/28866577/Notes_-Amino-Acids-and-Proteins_ (accessed in 2006).

Dodds, J.H. and Roberts, L.W. (1985). *Experiments in Plant Tissue Culture*, 2nd Ed. Cambridge: Cambridge University Press.

Dougherty, M.J. and Arnold, F.H. (2009). Directed evolution: New parts and optimized function. *Curr Opin Biotechnol* 20(4): 486–491.

Editorial (2011). Method of the Year 2010 (With the capacity to control cellular behaviors using light and genetically encoded light-sensitive proteins, optogenetics has opened new doors for experimentation across biological fields.) *Nature Methods* 8: 1.

Editorial (2012). Method of the Year 2011 (The ability to introduce targeted, tailored changes into the genomes of several species will make it feasible to ask more precise biological questions.) *Nature Methods* 9: 1.

Editorial (2013). Method of the Year 2012 (New method and tool developments are helping bring targeted proteome analysis technologies to a broader array of biologists.) *Nature Methods* 10: 1.

Editorial (2014a). Genome editing for all. *Nature Biotech* 32: 295.

Editorial (2014b). Method of the Year 2013 (Methods to sequence the DNA and RNA of single cells are poised to transform many areas of biology and medicine.) *Nature Methods* 11: 1.

Fire, A., Albertson, D., Harrison, S.W., and Moerman, D.G. (1991). Production of antisense RNA leads to effective and specific inhibition of gene expression in *C. elegans* muscle. *Development* 113: 503–514.

Fire, A., Xu, S.Q., Montgomery, M.K., Kostas, S.A., Driver, S.E., and Mello, C.C. (1998). Potent and specific genetic interference by double-stranded RNA in *Caenorhabditis elegans. Nature* 391: 806–811.

Gautier, A., Gauron, C., Volovitch, M., Bensimon, D., Julien, L., and Vriz, S. (2014). How to control proteins with light in living systems. *Nat Chem Biol* 10(7): 533–541.

Gehlenborg, N., O'Donoghue, S.I., Baliga, N.S., Goesmann, A., Hibbs, M.A., Kitano, H., Kohlbacher, O., Neuweger, H., Schneider, R., Tenenbaum, D., and Gavin, A.-C. (2010). Visualization of omics data for systems biology. *Nat Methods* Suppl. 7(3s): S56–S68.

Gelvin, S.B. (2003). *Agrobacterium*-mediated plant transformation: The biology behind the "gene-jockeying" tool. *Microbiol Mol Biol Rev* 67(1): 16–37.

Gerits, A. and Vanduffel, W. (2013). Optogenetics in primates: A shining future? *Trends Genet* 29(7): 403–411.

Goshen, I. (2014). The optogenetic revolution in memory research. *Trends Neurosci*, doi: 10.1016/j.tins.2014.06.002.

Hartl, D.L., Freifelder, D., and Snyder, L.A. (1988). *Basic Genetics*. Boston: Jones & Bartlett Publishers.

Harvey, A.J., Speksnijder, G., Baugh, L.R., Morris, J.A., and Ivarie, R. (2002). Expression of exogenous protein in the egg white of transgenic chickens. *Nat Biotech* 19: 396–399.

Howard, K. (2003). First Parkinson gene therapy trial launches. *Nat Biotech* 21(10): 1117–1118.

Howe, C. (2007). *Gene Cloning and Manipulation*. UK: Cambridge University Press.

The International Stem Cell Initiative (2007). Characterization of human embryonic stem cell lines by the International Stem Cell Initiative. *Nature Biotech* 25(7): 803–816.

Karimi, M., Inze, D., and Depicker, A. (2002). GATEWAY™ vectors for *Agrobacterium*-mediated plant transformation. *Trends Plant Sci* 7(5): 193–195.

Kawamura, T., Suzuki, J., Wang, Y.V., Menendez, S., Morera, L.B., Raya, A., Wahl, G.M., and Izpisua Belmonte, J.C. (2009). Linking the p53 tumor suppressor pathway to somatic cell reprogramming. *Nature* 460: 1140–1144.

Kelly, T.J. Jr. and Smith, H.O. (1970). A restriction enzyme from *Hemophilus influenza* II. *J Mol Biol* 51(2): 393–409.

Kling, J. (2009). First U.S. approval for a transgenic animal drug. *Nature Biotech* 27(4): 302–304.

Knight, R.D., Freeland, S.J., and Landweber, L.F. (2001). Rewiring the keyboard: Evolvability of the genetic code. *Nat Rev Genet* 2: 49–58.

Knopfel, T., Lin, M.Z., Levskaya, A., Tian, L., Lin, J.Y., and Boyden, E. (2010). Toward the second generation of optogenetic tools. *J Neurosci* 30(45): 14998–15004.

Koide, S. (2009). Generation of new protein functions by nonhomologous combinations and rearrangements of domains and modules. *Curr Opin Biotechnol* 20(4): 398–404.

Lai, L., Kang, J.X., Li, R., Wang, J., Witt, W.T., Yong, H.Y., Hao, Y., Wax, D.M., Murphy, C.N., Rieke, A., Samuel, M., Linville, M.L., Korte, S.W., Evans, R.W., Starzl, T.E., Prather, R.S., and Dai, Y. (2006). Generation of cloned transgenic pigs rich in omega-3 fatty acids. *Nat Biotech* 24(4): 435–436.

Larson, C. (2014). Genome editing: The experiment. *MIT Technology Review*, http://www.technologyreview.com/featuredstory/526511/genome-editing/ (accessed April 2014).

Leder, Z.P. and Nirenberg, M.W. (1964). RNA codewords and protein synthesis III. On the nucleotide sequence of a cysteine and a leucine RNA codeword. *PNAS* 52(6): 1521–1529.

Ledran, M.H., Krassowska, A., Armstrong, L., Dimmick, I., Renstrom, J., Lang, R., Yung, S., Santobanez-Coref, M., Dzierzak, E., Stojkovic, M., Oostendrop, R.A.J., Forrester, L., and Lako, M. (2008). Efficient hematopoietic differentiation of human embryonic stem cells on stromal cells derived from hematopoietic niches. *Cell Stem Cell* 3(1): 85–98.

Li, L. and Akashi, K. (2003). Unraveling the molecular components and genetic blueprints of stem cells. *Biotechniques* 35(6): 1233–1239.

Lima, S. and Miesenbock, G. (2005). Remote control of behavior through genetically targeted photostimulation of neurons. *Cell* 121(1): 141–152.

Liou, M.R., Huang, Y.W., Hu, C.C., Lin, N.S., and Hsu, Y.H. (2014). A dual gene-silencing vector system for monocot and dicot plants. *Plant Biotechnol* 12(3): 330–343.

Liu, C.C., Mack, A.V., Tsao, M.L., Mills, J.H., Lee, H.S., Choe, H., Farzan, M., Schultz, P.G., and Smider, V.V. (2008). Protein evolution with an expanded genetic code. *Proc Natl Acad Sci* 105(46): 17688–17693.

Liu, D.R. and Schultz, P.G. (1999). Progress toward the evolution of an organism with an expanded genetic code. *Proc Natl Acad Sci* 96(9): 4780–4785.

Liu, W., Yuan, J.S., and Stewart, Jr. C.N. (2013). Advanced genetic tools for plant biotechnology. *Nat Rev Genet* 14: 781–793.

Lonquespee, R., Fleron, M., Pottier, C., Quesada-Calvo, F., Meuwis, M.A., Maiwir, D., Smargiasso, N., Mazzucchelli, G., De Pauw-Gillet, M.C., Delvenne, P., and De Pauw, E. (2014). Tissue proteomics for the next decade? Towards a molecular dimension in histology. *Omics* 18(9): 539–552.

Mack, G.S. (2011). ReNeuron and StemCells get green light for neural stem cell trials. *Nat Biotech* 29(2): 95–97.

Madsen, O.D. and Serup, P. (2006). Towards cell therapy for diabetes. *Nat Biotech* 24(12): 1481–1483.

Malyshev, D.A., Dhami, K., Lavergne, T., Chen, T., Dai, N., Foster, J.M., Correa, I.R. Jr., and Romesberg, F.E. (2014). A semi-synthetic organism with an expanded genetic alphabet. *Nature* 509(7500): 385–388.

Mardis, E.R. (2007). The impact of next generation sequencing technology on genetics. *Trends in Genet* 24(3): 133–141.

Marshall, A. (2009). 13.3 million farmers cultivate GM crops. *Nat Biotech* 27(3): 221.

Matsumoto, K., Isagawa, T., Nishimura, T., Ogaeri, T., Eto, K., Miyazaki, S., Miyazaki, J., Aburatani, H., Nakauchi, H., and Ema, H. (2009). Stepwise development of hematopoietic stem cells from embryonic stem cells. *PLoS One* 4(3): e4820.

Matthaei, H.J., Jones, O.W., Martin, R.G., and Nirenberg, M.W. (1962). Characteristics and composition of RNA coding units. *PNAS* 48(4): 666–677.

Meselson, M. and Yuan, R. (1968). DNA restriction enzyme from *E. coli*. *Nature* 217: 1110–1114.

Metzker, M.L. (2010). Sequencing technologies—The next generation. *Nat Genet* 11: 31–46.

Nakagawa, M., Koyanagi, M., Tanabe, K., Takahashi, K., Ichisaka, T., Aoi, T., Okita, K., Mochiduki, Y., Takizawa, N., and Yamanaka, S. (2008). Generation of induced pluripotent stem cells without Myc from mouse and human. *Nat Biotech* 26(1): 101–106.

Nair, A.J. (2008). *Introduction to Biotechnology and Genetic Engineering*. Hingham, MA: Infinity Science Press.

Nelson, D.L. and Cox, M.M. (2008). *Lehninger Principles of Biochemistry*, 5th Ed. New York: W.H. Freeman.

Niiler, E. (2000). FDA, researchers consider first transgenic fish. *Nature Biotech* 18: 143.

Ogita, S., Uefuji H., Yamaguchi, Y., Koizumi, N., and Sano, H. (2003). RNA interference: Producing decaffeinated coffee plants. *Nature* 423: 823.

Okita, K., Ichisaka, T., and Yamanaka, S. (2007). Generation of germline-competent induced pluripotent stem cells. *Nature* 448: 313–317.

Ollivier, L. (2013). A comment on Seralini G.-E. et al., "Long-term toxicity of a Roundup herbicide and a Roundup-tolerant genetically modified maize" in *Food Chem Toxicol* 2012. *Food Chem Toxicol* 53: 458.

Palmiter, R.D. and Brinster, R.L. (1986). Germ line transformation of mice. *Ann Rev Genet* 20: 465–499.

Palmiter, R.D., Brinster, R.L., Hammer, R.E., Trumbauer, M.E., Rosenfeld, M.G., Birnberg, N.C., and Evans, R.M. (1982). Dramatic growth of mice that develop from eggs microinjected with metallothionein growth hormone fusion genes. *Nature* 300: 611–615.

Plumer, B. (2013). We've covered the world in pesticides. Is that a problem? *The Washington Post* Wonkblog, September 13. http://www.washingtonpost.com/wonkblog/wp/2013/08/18/the-world-uses-billions-of-pounds-of-pesticides-each-year-is-that-a-problem/.

Primrose, S.B. and Twyman, R.M. (2006). *Principles of Gene Manipulation and Genomics*, 7th Ed. Maiden, MA: Blackwell Pub.

Puchta, H. and Fauser, F. (2013). Gene targeting in plants: 25 years later. *Int J Dev Biol* 57: 629–637.

Purkayashta, A. and Dasgupta, I. (2009). Virus-induced gene silencing: A versatile tool for discovery of gene functions in plants. *Plant Physiol Biochem* 47: 967–976.

Redden, H., Morse, N., and Alper, H.S. (2014). The synthetic biology toolbox for tuning gene expression in yeast. *FEMS Yeast Res*, doi: 10.1111/1567-1364.12188.

Reddington, S., Watson, P., Rizkallah, P., Tippmann, E., and Jones, D.D. (2013). Genetically encoded phenyl azide chemistry: New uses and ideas for classical biochemistry. *Biochem Soc Trans* 41(5): 1177–1182.

Reece, R. (2004). *Analysis of Genes and Genomes*. England: John Wiley & Sons.

Richards, M., Tan, S.P., Tan, J.H., Chan, W.K., and Bongso, A. (2004). The transcriptome profile of human embryonic stem cells as defined by SAGE. *Stem Cells* 22(1): 51–64.

Rojahn, S.Y. (2014). Broad Institute gets patent on revolutionary gene-editing method. *MIT Technology Review*, April 16. http://www.technologyreview.com/view/526726/broad-institute-gets-patent-on-revolutionary-gene-editing-method/.

Romero, P.A. and Arnold, F.H. (2009). Exploring protein fitness landscapes by directed evolution. *Nat Rev Mol Cell Biol* 10(12): 866–876.

Sanders, D., Kamoun, S., Williams, B., and Festing, M. (2013). Letter to the editor: Response to paper by Seralini et al. on genetically modified maize. *Food Chem Toxicol* 53: 450–453.

Schuster, S.C. (2008). Next-generation sequencing transforms today's biology. *Nature Methods* 5(1): 16–18.

Seralini, G.E., Clair, E., Mesnage, R., Gress, S., Defarge, N., Malatesta, M., Hennequin, D., and Spiroux de Vendomois, J. (2012). Long-term toxicity of a Roundup herbicide and a Roundup-tolerant genetically modified maize. *Food Chem Toxicol* 50: 4221–4231 (Retracted).

Seralini, G.E., Mesnage, R., Defarge, N., Gress, S., Hennequin, D., Clair, E., Malatesta, M., and Spiroux de Vendomois, J. (2013). Answers to critics: Why there is a long-term toxicity due to a Roundup-tolerant genetically modified maize and to a Roundup herbicide. *Food Chem Toxicol* 53: 476–483.

Shaner, N.C., Steinbach, P.A., and Tsien, R.Y. (2005) A guide to choosing fluorescent proteins. *Nature Methods* 2(12): 905–909.

Sheridan, C. (2007). Positive clinical data in Parkinson's and ischemia buoy gene therapy. *Nat Biotech* 25(8): 823-824.

Sheridan, C. (2011). Gene therapy finds its niche. *Nature Biotech* 29: 121–128.

Siuti, P., Yazbek, J., and Lu, T.K. (2014). Engineering gene circuits that compute and remember. *Nat Protocols* 9(6): 1292–1300.

Smith, C.J.S., Watson, C.F., Bird, C.R., Morris, P.C., Schuh, W., and Grierson, D. (1988). Antisense RNA inhibition of polygalacturonase gene expression in transgenic tomatoes. *Nature* 334: 724–726.

Smith, H.O. and Wilcox, K.W. (1970). A restriction enzyme from *Hemophilus influenza* I. Purification and general properties. *J Mol Biol* 51(2): 379–391.

Subramaniam, S. (1998). The Biology Workbench—A seamless database and analysis environment for the biologist. *Proteins* 32, 1–2.

Takahashi, K. and Yamanaka, S. (2006). Induction of pluripotent stem cells from mouse embryonic and adult fibroblast cultures by defined factors. 126(4): 663–676.

Tanaka, K.F., Matsui, K., Sasaki, T., Sano, H., Sugio, S., Fan, K., Hen, R., Nakai, J., Yanagawa, Y., Hasuwa, H., Okabe, M., Deisseroth, K., Ikenaka, K., and Yamanaka, A. (2012). Expanding the repertoire of optogenetically targeted cells with an enhanced gene expression system. *Cell Rep* 2(2): 397–406.

Tsien, R.Y. (1998). The green fluorescent protein. *Annu Rev Biochem* 67: 509–544.

Valton, J., Cabaniols, J.P., Galetto, R., Delacote, F., Duhamel, M., Paris, S., Blanchard, D.A., Lebuhotel, C., Thomas, S., Moriceau, S., Demirdjian, R., Letort, G., Jacquet, A., Gariboldi, A., Rolland, S., Daboussi, F., Juillerat, A., Bertonati, C., Duclert, A., and Duchateau, P. (2014). Efficient strategies for TALEN-mediated genome editing in mammalian cell lines. *Methods*, http://dx.doi.org/10.1016/j.ymeth.2014.06.013.

Van Berkel, P.H.C., Welling, M.M., Geerts, M., van Veen, H.A., Ravensbergen, B., Salahaddine, M., Pauwels, E.K.J., Pieper, F., Nuijens, J.H., and Nibbering, P.H. (2002). Large-scale production of recombinant human lactoferrin in the milk of transgenic cows. *Nature Biotech* 20: 484–487.

Van Keuren, M.L., Gavrilina, G.B., Filipiak, W.E., Zeidler, M.G., and Saunders, T.L. (2009). Generating transgenic mice from bacterial artificial chromosomes: Transgenesis efficiency, integration and expression outcomes. *Transgenic Res* 18(5): 769–785.

Wang, H., Yang, H., Shivalila, C.S., Dawlaty, M.M., Cheng, A.W., Zhang, F., and Jaenisch, R. (2013). One-step generation of mice carrying mutations in multiple genes by CRISPR/Cas-mediated genome engineering. *Cell* 153: 1–9.

Wang, L., Menendez, P., Shojaei, F., Li, L., Mazurier, F., Dick, J.E., Cerdan, C., Levac, K., and Bhatia, M. (2005). Generation of repopulating cells from human embryonic stem cells independent of HOXB4 expression. *J Exp Med* 201(10): 1603–1614.

Wang, L. and Schultz, P.G. (2002). Expanding the genetic code. *Chem Commun* 1: 1–11.

Wang, Y., Yates, F., Naveiras, O., Ernst, P., and Daley, G.Q. (2005). Embryonic stem cell-derived hematopoietic stem cells. *Proc Natl Acad Sci* 102(52): 19081–19086.

Watson, J.D., Baker, T.A., Bell, S.P., Gann, A., Levine, M., and Losick, R. (2008). *Molecular Biology of the Gene*. New York: Pearson, Benjamin Cummings, CSHL Press.

Watson, J.D. and Crick, F.H.C. (1953). A structure for deoxyribose nucleic acid. *Nature* (3)171: 737–738.

Wilmut, I., Sullivan, G., and Taylor, J. (2009). A decade of progress since the birth of Dolly. *Reprod Fertil Dev* 21(1): 95–100.

Wilson, S.A., Cummings, E.M., and Roberts, S.C. (2014). Multi-scale engineering of plant cell cultures for promotion of specialized metabolism. *Curr Opin Biotech* 29C: 163–170.

Yang, H., Wang, H., and Jaenisch, R. (2014). Generating genetically modified mice using CRISPR/Cas-mediated genome editing. *Nat Protocols* 9(8): 1956–1968.

Ye, H. and Fussenegger, M. (2014). Synthetic therapeutic gene circuits in mammalian cells. *FEBS Lett* 588(15): 2537–2544.

Ye, X., Al-Babili, S., Kloti, A., Zhang, J., Lucca, P., Beyer, P., and Potrykus, I. (2000). Engineering the provitamin A (beta-carotene) biosynthetic pathway into carotenoid-free rice endosperm. *Science* 287: 303–305.

Yuan, J., O'Donoghue, P., Ambrogelly, A., Gundllapalli, S., Sherrer, R.L., Palioura, S., Simonovic, M., and Söll, D. (2010). Distinct genetic code expansion strategies for selenocysteine and pyrrolysine are reflected in different aminoacyl-tRNA formation systems. *FEBS Lett* 584(2): 342–349.

Zemelman, B.V., Lee, G.A., Ng, M., and Miesenbock, G. (2002). Selective photostimulation of genetically chARGed neurons. *Neuron* 33(1): 15–22.

Zemelman, B.V., Nesnas, N., Lee, G.A., and Miesenbock, G. (2003). Photochemical gating of heterologous ion channels: Remote control over genetically designated populations of neurons. *Proc Natl Acad Sci* 100(3): 1352–1357.

Zhang, Y., Baranov, P.V., Atkins, J.F., and Gladyshev, V.N. (2005). Pyrrolysine and selecocysteine use dissimilar coding strategies. *J Biol Chem* 280: 20740–20751.

Glossary

2D gel electrophoresis: Two-dimensional gel electrophoresis, generally used to study proteins, and involves separation of proteins into two dimensions; in the first dimension, separation based on isoelectric points of proteins, and in the second dimension, separation based on size of denatured proteins.

absorbed dose: A measure of radiation energy that is absorbed by unit mass, commonly used for organic matter (such as humans), and is important for estimating radiological protection; however, it does not directly indicate health effects on humans.

affinity: Degree of interaction between two molecules, such as receptor–ligand, antigen–antibody, or enzyme–substrate.

affinity purification: (Usually chromatographic) separation of proteins based on their affinity toward antibodies or other molecules (such as cations).

alkaline phosphatase: An enzyme that removes phosphate groups from substrate molecules preferably in alkaline conditions.

allele: One of the possible variations of the same gene or locus.

allogeneic: From genetically different sources.

antibiotic screening: Screening of transformation using antibiotic selection marker.

antibody: An immunoglobulin produced by the immune system to recognize foreign molecules or antigens.

antigen: Short for *anti*body *gen*erator, it is used for any molecule that provokes an immune reaction in the organism.

antisense RNA: A single-stranded RNA molecule that is complementary to the "sense" message, that is, the messenger RNA (mRNA). Sometimes abbreviated as *asRNA*.

autologous: From a genetically identical source (i.e., the patient himself or herself).

BAC: See *bacterial artificial chromosome.*

bacterial artificial chromosome: (Also called *BAC*) an F plasmid-based vector for cloning large inserts, commonly used for sequencing projects.

bacteriophage: A virus that infects bacteria. Also called a *phage.*

bait plasmid: A (yeast) two-hybrid plasmid, where the coding sequence for the protein of interest is fused to that of GAL4 DNA-binding domain and used as "bait."

Becquerel: The SI (International System of Units) unit of radioactivity, defined as one disintegration per second ($1\ s^{-1}$). Abbreviated as *Bq.*

bioinformatics: The study of large-scale biological data with the help of computer science, mathematics, statistics, and computational methods.

biolistic: A method for DNA delivery into cells that are otherwise difficult to transform. It involves gold or tungsten microparticles coated with DNA ("bullets"), which are then accelerated to a high velocity so as to penetrate even cell walls. Also called *gene gun* or *particle bombardment.*

bipotential: Stem cells or progenitors that can only produce two types of cells; having two developmental potentials.

blue-white screening: A rapid screening strategy for (bacterial) transformation, based on the *lac* operon principle; when grown in the presence of X-gal, recombinant bacteria will appear white, nonrecombinants will appear blue.

blunt end: The base-paired end of a double-stranded DNA molecule.

Bq (see *Becquerel*): The SI unit of radioactivity, corresponding to one nuclear decay per second. Also called *s–1*.

Bt: *Bacillus thuringiensis*, a biological pesticide microorganism which produces endotoxins that function as natural insecticides.

Bt cotton: The genetically modified cotton produced and patented by the Monsanto company, which is genetically modified to express the endotoxin genes from the microorganism *B. thuringiensis* as a method of pest control.

callus: Unorganized cell mass in plants, usually of parenchymal origin, which are partially undifferentiated, rapidly dividing cells.

cDNA: Also called *complementary DNA*, cDNA is a DNA molecule that is reverse transcribed using mRNA as a template (hence is "complementary" to mRNA). In eukaryotes, because mRNA is processed and introns are removed, cDNA corresponds to the "intronless" message.

cDNA library: A compilation of cDNA fragments corresponding to most of the mRNAs expressed in that cell or tissue, cloned into an appropriate library vector and host organism.

cell culture: A technique to grow, propagate, and manipulate live cells under defined physical and chemical conditions (i.e., pH, temperature, medium, and supplements).

centromere: A highly specialized region of chromatin where the two sister chromatids are linked, and which serves as the site of assembly for the kinetochore region of the chromosomes; may have variety of positions on the chromosome.

ChIP: See *chromatin immunoprecipitation*.

chromatin immunoprecipitation: Also called *ChIP* for short, it is a technique used to analyze the interaction between proteins and chromatin, using antibodies to precipitate the protein under investigation, followed by analysis of chromatin DNA with PCR.

cloning: The process of creating genetically identical copies (animal cloning); the technique used to create copies of (and amplify) fragments of target DNA using a vector and a host organism, such as bacteria (biotechnology).

co-IP: See *co-immunoprecipitation.*

codon: Nucleotide triplicates within coding regions of genes that correspond to an amino acid in the translated protein.

Co-Immunoprecipitation: Also called *Co-IP* for short.

colony PCR: A rapid screening strategy for transformation, based on PCR amplification of a desired target gene and the vector (using specific primers for each) directly from the colonies after a brief lysis process.

competent: Cells (typically bacterial) that are specifically processed to enable them to take up free DNA from the environment.

complementary DNA: See *cDNA*.

conditional knockout: A special type of *knockout* (see below), where the genetic modification is not done in the whole cell or organism, but is restricted to special cells or tissues using tissue-specific promoters, or is restricted to special conditions (such as specific temperatures, in the case of temperature-sensitive mutations or promoters, or the presence or absence of ligands in the case of inducible promoters, etc.).

constitutive expression: Continuous expression of a gene, regardless of cell type, conditions, or ligands, et cetera, present.

cosmid: A type of artificially engineered cloning vector (commonly for library construction), which includes elements from phade genome (the "cos" sites) and some features of plasmids (the "-mid"), thereby allowing cloning of larger size DNA fragments.

cre-lox system: A site-specific recombination system used extensively in cloning and transgenic technologies, based on the Cre enzyme of bacteriophage P1, which recognizes the *Lox* sites; the system can be used to insert or remove target fragments that have *loxP* sites on either end (a similar system is the Flp/Frt system).

CRISPR/Cas: A so-called prokaryotic immunity system, composed of CRISPRs (clustered regularly interspaced short palindromic repeats; remnants of previous viral exposures) and the related Cas protein. The guide RNA and endonuclease combination can "cleave" the DNA, which has been modified as a genome editing tool for introducing double-stranded breaks to desired genomic locations.

crown gall disease: A plant "tumor" caused by the bacterium *Agrobacterium tumafaciens.*

Curie (Ci): The SI (International System of Units) unit of radioactivity, defined as 3.7×10^{10} disintegration per second (hence, 3.7×10^{10} Bq). Abbreviated *Ci.*

deep sequencing: A recent advance in DNA sequencing technology, where the DNA fragment in question is sequenced (or "read") a large number of times, aimed, for example, at overcoming errors in genome sequencing projects; the "depth" refers to the number of times a given nucleotide in the target DNA is "read."

deoxyribonucleic acid: See *DNA*.

diagnostic digestion: A screening method for transformation, based on restriction digestion of the fragment or insert cloned inside a vector.

dicotyledon: A group of flowering plants that have characteristic features such as two embryonic leaves, which gives this group its name. Also known as *dicots.*

DNA: Deoxyribonucleic acid, a double-stranded polymer of a nucleotide unit, which consists of a deoxyribose sugar, a phosphate group, and a nitrogen-containing base; functions as the hereditary molecule and abbreviated as DNA.

DNA fingerprinting: A method that distinguishes each individual's DNA with relatively high significance, by analyzing highly variable repeats on the human genome that is unique to each individual (just like a fingerprint, hence the name).

DNA footprinting: A method used to identify DNA motif sequences that a protein recognizes and binds to.

DNA library: A compilation of DNA fragments, cloned into an appropriate library vector and host organism with the purpose of storage and propagation.

DNA sequencing: The method of determining the sequence of nucleotides in a DNA strand.

electrophoretic mobility shift assay: A method to study protein—nucleic acid interactions on nondenaturing gel, using differential mobilities of protein-bound and protein-unbound DNA molecules. Also known as *EMSA*, gel shift assay, gel retardation assay, and band-shift assay.

ELISA: See *enzyme-linked immunosorbent assay.*

EMSA: See *electrophoretic mobility shift assay.*

endonuclease: Enzyme that cuts from within a nucleic acid sequence, by cleaving phosphodiester bonds between two nucleotides.

enzyme-linked immunosorbent assay: Abbreviated *ELISA*, also known as *enzyme immunoassay (EIA)*, this is a biochemical assay method where an unknown amount of test antigen is immobilized on a surface, after which an antibody specific to that antigen is allowed to bind to the antigen. The enzyme that is conjugated to the specific antibody is then used to quantify the amount of antigen present on the surface, most commonly through colorimetric detection.

epigenetics: The study of hereditary properties of the genome that are not directly related to the DNA sequence.

epitope: A part of an antigen that can be recognized by an antibody and generate an immune response.

eugenics: A branch of science that aimed at improving the qualities of the human race through genetic breeding.

expression vector: A vector DNA that is specifically used for the cloning and expression of a coding sequence of a given gene in a particular host cell or organism.

Flavr Savr tomato: The first commercially available genetically modified crop plant, through which ripening was slowed down by antisense technology.

fluorescence resonance energy transfer: A technique used to study interaction of two proteins (within several nanometers) through transfer of resonance energy between two fluorophores attached to the proteins under investigation. Abbreviated as *FRET.*

fluorophore: A molecule that can emit fluorescence when excited at a particular wavelength.

FRET: See *fluorescence resonance energy transfer.*

gene: A functional unit of heredity. A region of DNA that, when expressed, affects one or more traits in a cell or organism.

gene gun: See *biolistic.*

gene therapy: A therapeutic approach to diseases using genetic engineering tools. Ideally it involves replacement of a mutated or dysfunctional gene with a healthy, normal version, although other approaches are available.

gene trapping: A large-scale "knockout" approach, where a candidate gene is screened through random insertional mutagenesis of the genome in animals.

Genentech: A pioneering biotechnology company that has produced several commercial biotech products since 1976, founded originally by Swanson and Boyer, (and now is a subsidiary of Roche).

genetic engineering: The use of molecular biology tools and techniques to change an organism's genome by introducing a novel DNA.

genetics: The study of heredity.

genome: The entire genetic material of an organism.

genomic DNA library: A compilation of genomic fragments of a target organism, cloned into an appropriate library vector and host organism.

genomics: The study of the complete genome of an organism.

genotype: The genetic makeup of a cell or organism with respect to a particular trait(s).

glutathione S-transferase: A large group of detoxifying enzymes that can conjugate the reduced form of glutathione to a given substrate. Abbreviated as *GST.*

Golden Rice: A type of rice that has been genetically modified to express beta-carotene producing enzymes from other organisms.

Gray: A measure of the absorbed dose of radiation (Gy).

GST: See *glutathione S-transferase.*

half-life: (*In radioactivity*) the time it takes for the radioactive isotopes in a sample to decay by half.

heavy isotope: The isotope which is stable, yet has more neutrons than the commonly found isotope (ex. ^{12}C is the common isotope, ^{13}C is the "heavy" yet stable isotope, while ^{14}C is the radioactive isotope).

heredity: The collection of all processes whereby offspring acquire certain traits or characteristics from parents.

high copy number plasmid: Plasmid that has relatively relaxed control of replication initiation, resulting in 10–100 copies of plasmid per cell.

His tag: A protein tag used for purification of proteins, usually consisting of 6 to 14 histidine amino acids (hence sometimes also called a *polyhistidine tag*).

homologous recombination: A type of recombination between two DNA helices of extensive overall homology (hence the name), seen in crossing over of meiosis, double-stranded break repair, or recovery of stalled replication forks.

homunculus: A small representation of a human body (*pl.* homunculi).

Human Genome Project: An international effort to sequence the entire human genome, which was initiated in 1990 and successfully completed in 2003.

hybridoma: A hybrid of antibody producing B cell with myeloma for the purpose of large-scale production of antibodies.

immunocytochemistry: A technique used to analyze subcellular localization of proteins or other molecules within cells, using a combination of primary and secondary antibodies with a fluorophore conjugate.

immunofluorescence: A technique used to analyze subcellular localization of proteins or other molecules within cells, using a combination of primary and secondary antibodies with a fluorophore conjugate, as well as other direct fluorescent staining methods.

immunohistochemistry: A technique used to analyze subcellular localization of proteins or other molecules within a tissue, using a combination of primary and secondary antibodies with a fluorophore or chromophore conjugate.

immunoprecipitation: A technique that uses antigen antibody affinity for crude precipitation and purification of proteins. Also called *IP*.

in frame: Without disrupting the sequence of nucleotide triplets of a coding sequence, or the reading frame.

***in vitro*:** Outside the natural environment or organism.

***in vitro* transcription and translation:** Transcription and translation of a coding sequence in a cell-free system.

incompatible: (For plasmids) having the same origin of replication and thus not possible to maintain in the same organism simultaneously, in the absence of selection pressure.

induced pluripotent stem cells: Cells (usually somatic) that have been "induced" by genetic modification to resemble stem cells in terms of pluripotency and self-renewal. Abbreviated as iPSCs.

inducible expression: Gene expression system where transcription from a promoter can be regulated or "induced" by a physical or chemical agent.

insertional vector: A vector where a specific site in the middle is digested and the target sequence can be *inserted.*

internal ribosome entry site: *Cis*-acting DNA sequences in the middle of an mRNA sequence that can recruit ribosomal subunits to allow from an internal start codon. Abbreviated as *IRES.*

IP: See *immunoprecipitation.*

iPSC: See *induced pluripotent stem cells.*

IRES: See *internal ribosome entry site.*

isoschizomers: Restriction enzymes that recognize the same recognition motif and cleave from the same position, but are named differently due to different species they were isolated from.

isotope: An isotope has the same number of protons but different number of neutrons.

Joule: The SI unit of energy or work (J).

Klenow fragment: The large fragment of *E. coli* DNA polymerase I, which retains the 5′-to-3′ polymerase and 3′-to-5′-exonuclease activities.

knockout: A technique where the genome of either a cell (when creating a knockout cell) or an embryonic stem cell (or equivalent, when creating a knockout organism) is modified in such a way as to render a particular target gene nonfunctional (through deletion of one or more exons, or interruption of a gene by insertion of a marker gene, etc.) in a stable fashion (i.e., modification transmitted through generations).

ligase: (DNA ligase) A modifying enzyme that can *ligate*, or form a phosphodiester bond between, a 5′-phosphate group and a nearby free 3′-hydroxyl group in a DNA molecule.

lipidomics: The study of the complete set of lipids in a cell or organism.

lipofection: Liposome-mediated transfer of nucleic acids (or drugs) to living systems (such as cells, tissues, or animals).

low copy number plasmid: Plasmid that has stringent control of replication initiation, resulting in one to two copies of plasmid per cell.

marker: Any genetic or other way to "mark" a particular cell, tissue, or organism; (in cloning) a selectable marker such as an antibiotic resistance gene identifies cells that have taken up the recombinant DNA; in transgenic animals or plants, fluorescent markers or antibiotic resistance markers, et cetera, help identify the organisms that have stably incorporated the transgene.

MAT: See *mating type.*

mating type: The genes that are responsible for types of sexual reproduction in the haploid life cycle of yeast. Also called *MAT.*

MCS: See *multiple cloning site.*

meganuclease: Microbial endonucleases with rather large (~18–20 bp long) recognition motifs.

metabolomics: The study of the complete set of metabolites in a cell or organism.

methylase: DNA methylase, which adds methyl groups to DNA molecules at specific sites (such as adenine within a 5′-GATC-3′ in the case of *Dam* methylase).

methylation: Chemical linkage of a methyl group to a substrate (such as DNA or protein).

microarray: A method used for large-scale study of gene expression, polymorphisms, et cetera, using a microscale array of unique DNA sequences on a so-called DNA chip.

microRNA: Small noncoding RNA molecules, either expressed from microRNA genes or produced from introns, and capable of suppressing expression of target mRNAs.

miRNA: See *micro RNA*.

molecular biology: The study of biological systems at the molecular level, using concepts of chemistry, biology, and physics.

monoclonal antibody: An antibody that is highly (mono) specific to a given epitope antigen, derived from a single parent B cell *clone* (hence, mono-clonal).

monocotyledon: The type of flowering plants with only one cotyledon.

multiple cloning site: A short engineered sequence within a cloning or expression vector that contains a number of unique restriction enzyme recognition sequences not present in the base vector. Also called *polylinker*. Abbreviated as *MCS*.

multipotent: Stem cells or progenitors that can produce multiple types of cells within the same organ; more limited potential than pluripotent cells.

mutagenesis: Any process which causes a change in the DNA sequence.

neoschizomers: Restriction enzymes that recognize the same recognition motif but cleave from a different position within the sequence.

Northern blot: A method used to detect target RNA species within a mixture through hybridization to a probe.

nuclease protection assay: A method used to identify specific RNA molecules within a mixture through hybridization to a probe and digestion of nonhybridized RNA species by a ribonuclease.

oligopotent: Stem cells or progenitors that can produce a few types of cells within a lineage.

operon: A set of genes that function as a unit within the same metabolic pathway and is expressed from a single promoter.

optogenetics: Genetic manipulation of neurons, muscle, or other cell types to express a light-sensitive channel protein so as to control the cell function through light stimulation.

organ culture: A tissue culture method to grow part of or the entire organ *in vitro*.

organotypic culture: A tissue culture method to grow 3-dimensional assembly of various cell types within a tissue *in vitro*, resembling the target organ under study.

ori: See *origin of replication*.

origin of replication: A specific DNA sequence within a genome that is used to initiate DNA replication. Also called *ori*.

overhang: The single-stranded DNA stretch that "hangs over" after cleavage with a restriction enzyme with a staggered cut site.

palindrome: A specific DNA sequence motif that "reads" the same 5′-to-3′ on both strands of the sequence.

PCR: See *polymerase chain reaction*.

phage vector: Cloning vectors based on bacteriophages.

phagemid: Cloning vector that has features from phage vectors (usually phage origin of replication) as well as features from plasmid vectors.

pharm animal: Transgenic animals genetically engineered to produce pharmaceutically valuable therapeutic molecules or proteins.

plaque assay: A method commonly used to determine viral concentration or titer in terms of plaque-forming units.

plasmid: Small, circular, extrachromosomal, self-replicating DNA unit.

pluripotent: Stem cells or progenitors that can produce essentially all types of cells within an embryo.

polyclonal antibody: Antibody that is specific to an antigen, but derived from multiple parent B cell "clones" (hence poly-clonal, and usually recognizing different epitopes within the same antigen).

polymerase: An enzyme that generates nucleic acid polymers from nucleotide building blocks.

polymerase chain reaction: Abbreviated as *PCR*. It is a laboratory technique designed to amplify targeted DNA regions *in vitro*, using a process known as thermal cycling with heat-resistant DNA polymerases, primer oligonucleotides complementary to boundaries of target regions, deoxynucleotides and buffers in a so-called thermal cycler; uses 20–30 cycles of heat denaturation of template DNA (usually at around 94°C), annealing of primers (at around 45–60°C), and elongation (at around 72°C).

polymorphism: (*In genetics*) variations within a DNA sequence that are common within a population.

potency: The ability of a cell to differentiate into different cell types.

prey plasmid: A (yeast) two-hybrid plasmid, where the coding sequence for (all) potential/suspected partner(s) of the target protein is fused to that of the GAL4 activation domain.

primary cell culture: *In vitro* culture of cells directly derived from a tissue and dissociated into dispersed cells (usually and preferably of a single cell type).

primer: A short stretch of single-stranded DNA that acts as a starting point or a *primer* of DNA replication.

probe: A relatively long (in the range of 100–1000 bp) stretch of DNA or RNA that is radioactively or fluorescently labeled, and used for detection of a target.

promoter: A region of DNA upstream of a gene that can initiate transcription of that gene.

proteomics: The study of the complete set of proteins in a cell or organism.

protoplast: Plant cell that is devoid of its cell wall.

pull-down assay: A technique that uses enzyme–substrate or protein–metal, et cetera, affinity to study protein–protein interactions.

pulse-chase analysis: A method to study a dynamic molecular event (such as replication, translation, or a metabolic pathway) in a live cell, using a series of steps of exposing the cells to a radioactively labeled precursor followed by an unlabeled "cold" version of the precursor.

radioactive labeling: Incorporating a radioactive compound to the molecule to be analyzed or traced.

radioisotope: An unstable isotope that has an excess of neutrons (ex. ^{12}C is the common isotope, ^{13}C is "heavy" yet stable isotope, while ^{14}C is the radioactive isotope).

reading frame: One of six possible ways of "reading" any given DNA sequence, or dividing the sequence into triplet codons, for translation.

real-time PCR: A version of PCR reaction that is modified so as to use fluorescent dye molecules to optically monitor the reaction in real time.

recombinant: An organism that has acquired a new combination of alleles, different from either of its parents, or source strain.

recombinant DNA: A new combination of DNA, generated by genetic engineering from two different DNA molecules.

recombinase: An enzyme that catalyzes DNA recombination reaction.

recombination: A process by which either the same DNA molecule is "cut" and joined at different positions, or two different DNA molecules are combined (e.g., strand exchange of sister chromatids during meiosis, or crossing over).

replacement vector: A phage vector where a large portion of the phage genome has been replaced with a "stuffer" DNA; the target DNA insert will then be replaced with this stuffer during cloning.

reporter(s): See *reporter gene.*

reporter gene: A gene that is used to *report* either if a transgene has been taken up by a cell, or whether the gene of interest is expressed in a cell type, or where and when.

restriction endonuclease: See restriction enzyme.

restriction enzyme: Endonucleases that recognize specific DNA sequence motifs and cleave the sugar-phosphate backbone on DNA. Also called *restriction endonuclease.*

restriction fragment length polymorphism: A method devised to analyze genetic variations among samples, which are associated with variations in restriction enzyme recognition motifs within a DNA region. Abbreviated as *RFLP.*

restriction map: A map of all known restriction sites on a given DNA region, with the nucleotide positions with respect to each other.

reverse transcriptase: RNA-directed DNA polymerase of retroviruses, used for synthesis of cDNA from mRNA template.

reverse transcription: Synthesis of a DNA molecule from an RNA template, in molecular biology synthesis of cDNA from mRNA template.

RFLP: See *restriction fragment length polymorphism.*

ribonucleic acid: See *RNA.*

RNA: Ribonucleic acid, a single-stranded polymer of a nucleotide unit, which consists of a ribose sugar, a phosphate group, and a nitrogen-containing base; functions as messenger RNA, transfer RNA, ribosomal RNA, as well as other types of small RNA molecules.

RNA interference: The process where special RNA molecules act to inhibit gene expression. Abbreviated as *RNAi.*

Roundup Ready Soybean: A genetically modified soybean variety, produced and patented by Monsanto company, which is rendered resistant to herbicide glyphosate, called *Roundup.*

RT-PCR: Reverse transcription polymerase chain reaction, used to PCR-amplify expressed sequences after converting them from mRNA to cDNA.

screening: (In molecular cloning) a method to identify true recombinants among a group of transformed cells or organisms.

SDS-PAGE: A method for separating denatured proteins according to size. Short for sodium dodecyl sulphate poplyacrylamide gel electrophoresis.

selectable marker: See *marker.*

self-renewal: A process by which stem cells divide (symmetrically) to produce identical copies of themselves, generating a stem cell clone.

short hairpin RNA: A regulatory RNA with a secondary structure resembling a hairpin, which (upon processing) functions in RNA interference. Abbreviated as *shRNA.*

short interfering RNA: A short stretch of double-stranded RNA molecule (usually of 20–22 bp) that functions in RNA interference. Abbreviated as siRNA.

shRNA: See *short hairpin RNA.*

Sievert: An SI unit of ionizing radiation dose as a measure of health effects on humans (Sv).

siRNA: See *short interfering RNA.*

site-specific recombination: A subtype of recombination, where two different DNA molecules (which may or may not have any overall sequence homology) exchange strands through very short yet identical sequence motifs or sites (which are specifically recognized by recombinase enzymes; hence, site-specific).

somatic cell nuclear transfer: A technique used in animal cloning through transfer of the diploid nucleus of a somatic cell into an enucleated embryo. Abbreviated as SCNT.

Southern blot: A method used to detect target DNA species within a mixture through hybridization to a probe.

stable transfection: Stable and long-term introduction of foreign DNA into a cell or an organism (usually through integration of the foreign DNA into the host cell's genome).

star activity: Relaxation of the target motif recognition specificity of the restriction enzymes due to altered and suboptimal reaction conditions.

stem cell: An undifferentiated cell that is capable of unlimited self-renewal capacity as well as the potential to differentiate into multiple cell types.

sticky end: The non-base-paired end of a double-stranded DNA molecule, where a short stretch is "sticking" out.

strain: A genetic variant of a microorganism, cell, plant, or animal used in a laboratory, with distinct and unique features, yet not categorized as an entirely different breed or variety.

syngeneic: Genetically identical individuals of the same species, particularly with respect to antigenic responses or immunological reactions.

synthetic biology: The interdisciplinary science that combines chemistry, biology, and engineering among many others, to combine engineering principles with biological systems so as to create novel biological functions, products, or devices.

systems biology: A holistic approach to understanding biological systems, interactions, networks, and pathways.

tag: (1) Short peptide sequences that are added to a protein through genetic engineering, used to identify, track, or purify a protein; (2) chromophore or fluorophore, et cetera, compounds that are conjugated to antibodies or probes for tracing or analysis.

TALEN(s): See *transcription activator-like effector nucleases.*

telomere: Relatively long stretch of repetitive DNA sequences at either end of chromosomes.

teratocarcinoma: A type of germ cell tumor.

tet-ON: A type of inducible gene expression system, where the expression of the cloned gene is turned "ON" upon administering the compound tetracycline, or rather its more stable analog doxycycline.

tet-OFF: A type of inducible gene expression system, where the expression of the cloned gene is turned "OFF" upon administering the compound tetracycline, or rather its more stable analog doxycycline.

Ti plasmid: The tumor-inducing (Ti) plasmid of the bacterium *Agrobacterium tumafeciens* that infects plants.

tissue culture: A technique to grow, propagate, and manipulate live cells or tissues under defined physical and chemical conditions (i.e., pH, temperature, medium, and supplements).

totipotent: Stem cells or progenitors that can produce all types of cells within an embryo as well as extra-embryonic cells.

transcription activator-like effector nucleases: (TALENs) Artificial restriction enzymes engineered to digest practically any target DNA sequence, making use of TAL effector DNA-binding domains tailored to recognize any desired sequence, and a DNA cleavage domain.

transcriptomics: The study of the complete set of transcripts (mRNAs) in a cell or organism.

transfection: Introduction of foreign DNA into cells, tissues, or organisms.

transfection efficiency: The percentage of cells in a population that have successfully been transfected with the foreign DNA.

transformation: Direct uptake of foreign DNA by competent microorganisms.

transformation efficiency: The percentage of microorganisms in a mixture that have successfully taken up the foreign DNA.

transgenic: An organism that has been genetically modified by a foreign DNA (the transgene) to achieve a novel property.

transient transfection: Short-term introduction of foreign DNA into a cell or an organism, expression of which usually lasts one to two days.

transposon: A sequence of DNA that can "jump," or change its position, within the genome, also called a *transposable element.*

unipotential: A progenitor that is not yet fully differentiated, but is nonetheless committed to produce only one type of cell; having only one developmental potential.

unnatural amino acids: Amino acids that are found in nature (as biosynthetic intermediates or other) and yet not used by the translational machinery, hence not naturally incorporated into proteins.

unnatural genetic alphabet: Laboratory-designed synthetic nucleotides that can form base pairs and are replicated and transcribed when incorporated into a DNA double helix.

vector: A vehicle DNA that is used to transfer foreign DNA into a host cell or organism.

Western blot: A method used to detect target protein species within a mixture through antigen-antibody recognition and signal detection.

xenogeneic: From a genetically different individual of the same species, hence immunologically incompatible.

YAC: See *yeast artificial chromosome.*

yeast artificial chromosome: Genetically engineered artificial chromosomes that carry key elements of yeast chromosomes. Abbreviated as *YAC.*

yeast two-hybrid: A technique used to analyze protein–protein interactions in living yeast cells.

zinc finger nuclease: Genetically engineered endonucleases that contain DNA-binding zinc finger domains and the catalytic domain of a restriction enzyme. Also called *ZFNs.*

Appendix A: DNA Techniques

As stated before, this is not a protocols book, and thus it is not intended to give a detailed technical background. However, this appendix on DNA techniques is simply intended to help the reader follow the main text where and when necessary. Here, we will only focus on major but not all of the molecular techniques used for the manipulation and analysis of DNA. Some advanced manipulation methods have been discussed in the main text.

A.1 DNA Gel Electrophoresis

The most common method used in DNA analysis is agarose gel electrophoresis of DNA. Agarose is a polysaccharide that forms a matrix, or gel, with nonuniform and relatively large pore-size gels, suitable for separation of DNA fragments of over 100 base pairs (for shorter fragments, other gel material, such as acrylamide, is commonly used). The negatively charged DNA molecule is thus "loaded" onto the gel, and is subjected to an electrical field, where the molecule moves toward the positively charged cathode. The shorter the DNA, the easier it is to pass through the agarose gel; hence, size separation of DNA molecules is possible (Figure A.1).

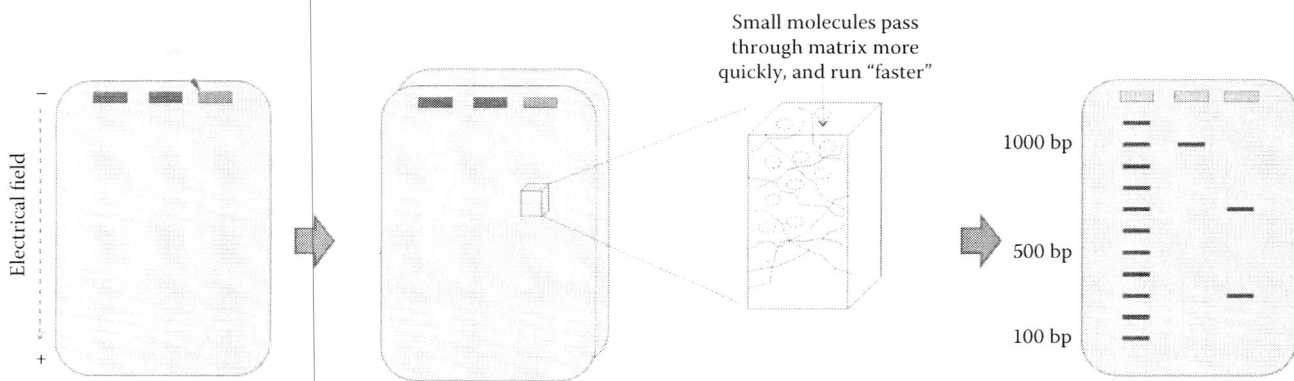

Figure A.1 A schematic depiction of DNA agarose gel electrophoresis. The DNA samples are loaded into the wells (far left panel) and subjected to an electrical field. DNA molecules travel through the agarose gel matrix that contains nonuniform pores (depicted as dashed circles); the shorter the DNA molecule, the faster it *runs* in the electrical field (middle panel). The size-separated DNA molecules are then visualized (by ethidium bromide or other methods) as bands (far right panel).

Although not discussed in the main text, it should be noted that capillary electrophoresis has become a common microfluidic tool used in much contemporary equipment, particularly for automation and high-throughput assays. The system is mainly composed of sample vials, a capillary tube filled with an electrolyte, and a power supply. When voltage is applied to the sample in the capillary tube, the sample solution moves into the capillary by a combination of capillary action, pressure, and electrical field, and "moves" toward the cathode. The analytes within the sample will migrate differently within the capillary, analyzed with the help of a computer at the end of the capillary, even when the sample size is very small (and not suitable for conventional agarose electrophoresis). This system gives capillary electrophoresis another huge advantage over agarose gel electrophoresis. Multiple capillaries can run in parallel, generating a system suitable for high-throughput analysis.

A.2 Nucleic Acid Blotting

Historically, perhaps the oldest blotting method that we know today was created by Edwin Southern, which he had devised so as to locate specific DNA sequences on gel. The **Southern blotting** method essentially involves the digestion of the (genomic, amplified, etc.) DNA with a restriction enzyme, giving a mixture of DNA fragments of different sizes, which produces a "smear" when run on agarose gel electrophoresis. The DNA thus separated in the gel is then denatured by an alkaline treatment so that the DNA can be dissociated into single strands that can then be hybridized to the probes. The DNA is then transferred from the agarose gel to a (nylon) membrane and cross-linked, that is, immobilized. This membrane is then treated with a (usually radiolabeled) probe, which recognizes and hybridizes with homologous sequences. The membrane is washed to get rid of excess probe and exposed to film or a phosphorimager cassette, after which the bands are visualized at the sites of hybridization (Figure A.2).

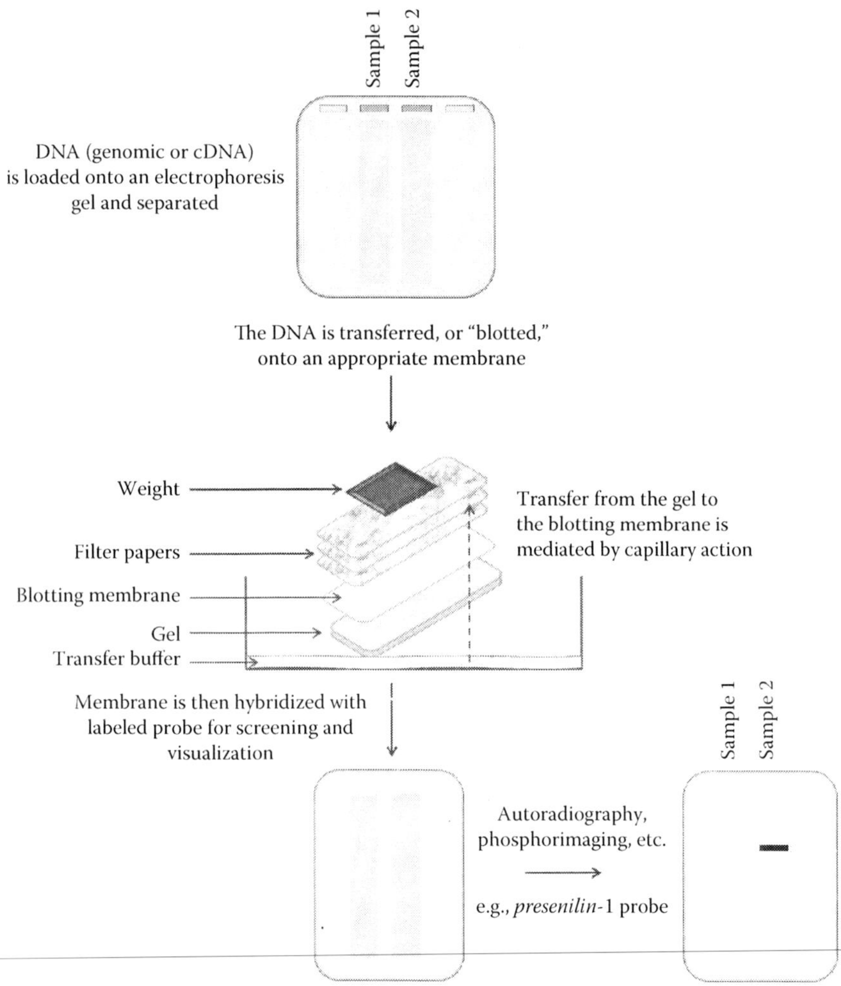

Figure A.2 A schematic depiction of Southern blotting. The DNA samples are loaded into the wells, and then *blotted* onto a membrane. The membrane is then hybridized with a labeled probe against the target DNA (for example, the presenilin-1 gene in Figure A1.2). After autoradiography or other methods of visualization, any DNA that hybridizes to the probe is seen as a *band* on the membrane.

A.3 Polymerase Chain Reaction

A polymerase chain reaction (PCR), developed by Kary Mullis in 1983, has been a routinely used method in almost every molecular biology laboratory for quite some time. Kary Mullis and Michael Smith were later awarded the Nobel Prize in Chemistry for this invention (10 years later in 1993). PCR entails amplifying a particular target DNA sequence *in vitro*, using heat-stable DNA polymerases. (Originally, *Taq* DNA polymerase from *Thermus aquaticus* was used but various other enzymes have been isolated from different bacteria and exploited commercially since then.)

In order to amplify a target sequence, one must have at least an idea about the sequence, since primers (short oligonucleotides bordering the target sequence) must be designed (Figure A.3).

These primers need to be as specific as possible for the target sequence (Figure A.3b); however, if, for example, only the protein sequence is known, then a coding sequence will be "predicted" using the genetic code table. Since multiple codons may code for the same amino acids, there will be many alternatives for the predicted coding sequence, and this will be reflected in the designing of so-called degenerate primers (Figure A.3c). Primers are typically around 18–20 bases in length, seldom going up to 30 bases, but they should have around 40 to 60% G-C content, and have matching annealing temperatures. These temperatures are estimated as roughly 10°C lower than the melting temperature, the simplest and quickest formula of which is $Tm = 2 \times (n_A + n_T) + 4 \times (n_G + n_C)$°C, where n is the number of the base indicated as subscript in the primer sequence. It is also a good idea to have one or two G or C residues at the end of the primer, so that 3′ ends bind strongly to the template, necessary for efficient elongation. It is also a good idea to check that the primers do not anneal to each other, forming primer dimers, and that they do not bind elsewhere in the template (for example, genomic

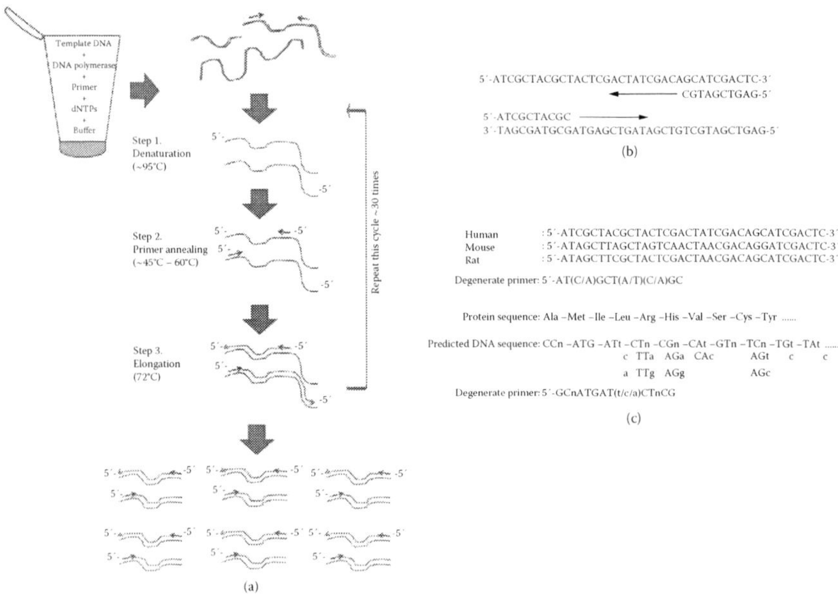

Figure A.3 A schematic depiction of PCR. (a) The DNA template, DNA polymerase, dNTPs, primers, and buffer are mixed in a reaction tube. In Step 1, template DNA in the tube is denatured by heat; in Step 2, the temperature is lowered so the primers can anneal to their specific targets; in Step 3, DNA polymerase in the tube elongates from the primers and amplifies the DNA. This cycle is repeated around 30 times, yielding millions of copies of target DNA region. (b) A gross simplification of primer design: primers are designed complementary to the borders of the target region, therefore the target sequence must be known (see specialty books or manuals for the details for designing primers). (c) If the exact sequence is not known (e.g., the protein sequence is known but the coding DNA sequence is not), or if the primer is designed to amplify the same DNA region from a number of different species (such as a human, mouse, and rat in the example, where some single nucleotide differences are possible), then a *degenerate* primer is to be designed. (Please note that the sequences and primers here are purely hypothetical, and similarly, the hypothetical primer designs shown in Figure A1.3 are not optimal.)

DNA); a BLAST (Basic Local Alignment Search Tool at http://blast.ncbi.nlm.nih.gov/Blast.cgi) search is usually pretty satisfactory for this purpose.

Once the primers are designed and a template is obtained, they will be incubated along with a DNA polymerase of choice, buffer, and dexoyribonucleotides, or dNTPs (Figure A.3a). In Step 1, template DNA will be denatured at a high temperature (usually 95°C), followed by Step 2, primer annealing. After the primers are allowed to anneal, DNA polymerase is set to elongate from the primers. This cycle is repeated around 30 times, generating millions of copies of the amplified region. These amplified regions can be visualized on an agarose gel, and the resulting DNA band can either be used for cloning, or sent for sequencing, and so forth, after purification.

A.4 Real-Time PCR

A more recent modification of the PCR amplification is the so-called real-time PCR, or quantitative PCR (qPCR), since it employs fluorescent dyes to monitor the amplification process in *real time*. There are many different versions of qPCR, either using a fluorescent DNA dye such as SYBR Green, which binds to double-stranded DNA (i.e., it is nonspecific), or using gene-specific probes that are fluorescently labeled. The principle behind the first strategy for qPCR is that since the fluorescent dye binds to double-stranded DNA, as the amount of DNA doubles up at each cycle (2^n), the amount of fluorescence is doubled, which can be measured automatically at the end of each cycle (hence, *real time*) and recorded. In the second strategy, since gene-specific probes will only bind to the target sequence, it is highly specific to the region that is amplified; the fluorescently labeled probe will bind to the amplified DNA at each cycle, increasing in intensity as the cycles progress, and as above, the fluorescence intensity is measured at the end of each cycle (again, real time). The slight difference here is that normally the probe has both a fluorescent dye and a *quencher*, preventing any fluorescence emission from the unbound probe. When the annealing step is reached, both the probe and the primers bind to the target sequence, and during elongation, the DNA polymerase that extends the sequence from the primer will reach the probe, which leads to the degradation of the 5′-quencher due to the 5′-3′ exonuclease activity of the polymerase. Release of the quencher from the probe during this elongation step will free the fluorescent dye of the probe, and emission will be detected and recorded. Other than these details, the qPCR reaction is pretty much the same as a normal PCR reaction. It should be noted that there are many different variations of this original probe-based qPCR strategy that we will not discuss here in detail.

A.5 DNA Sequencing

DNA sequencing was initially developed in the 1970s, and the original method that became routinely used for sequencing DNA was developed by Maxam and Gilbert (i.e., known as *Maxam–Gilbert sequencing*) and is based on chemical modification of DNA, followed by site-specific cleavage of this modified DNA. Since this method relies on the use of hazardous chemicals and radioactivity, it is hardly ever used on a routine basis in molecular biology laboratories; therefore, we will not go into the details of this outdated method. Instead, we will introduce the more recent, dideoxynucleotide-based Sanger method, or the chain-termination method, which is also the basic principle in many of the modern automated sequencing technologies.

In Sanger or dideoxy sequencing (also known as the *chain termination method*), dideoxynucleotides (ddNTPs) are used in addition to the normal deoxynucleotides (dNTPs): since dideoxynucleotides lack both 2′- and 3′-OH groups, 5′-to-3′ extension between 5′-phosphate and 3′-OH groups cannot take place, hence "terminating" the elongation of new DNA synthesis. Essentially, DNA synthesis takes place from a single-stranded template using a oligonucleotide primer (called the *sequencing primer*) and dNTPs. After a brief period of incubation this reaction is aliquotted into four different tubes, each of which contains a different labeled (radioactively or fluorescently) dideoxynucleotide terminator (Figure A.4). Fluorescently labeled ddNTPs are more commonly used in automated sequencers, where different fluorescence emissions are observed by a detector, base by base, giving the sequence of the template.

A.6 Next Generation Sequencing

Next generation sequencing, or alternatively high-throughput sequencing, actually dates back to the 1990s—to the *massively parallel signature sequencing*, developed by Lynx Therapeutics (founded by Sydney Brenner and Sam Eletr). The idea behind high-throughput technologies in general is to screen a large amount of data for a minimum price and in as short a time as possible. As such, it is not surprising that next-generation sequencers generate data that is equivalent to several hundred Sanger-type sequencers, operated by one person in one day (Mardis 2007; Schuster 2008). Recently, Roche Diagnostics released 454 pyrosequencing (also known as *sequencing by synthesis*) technologies, where DNA is captured by immobilization on beads and amplified inside water-oil emulsion. The machine uses many wells with picoliter volume to amplify from this template DNA, and uses luciferase for monitoring (the addition of new nucleotides results in a light signal); however, the cost per base that is sequenced is too high. Illumina, on the other hand, has developed another high-throughput method that uses reversible dye terminators (Metzker 2010). As these nucleotides are added to the DNA one at a time, the fluorescence of the nucleotides is recorded. Last, ABI's

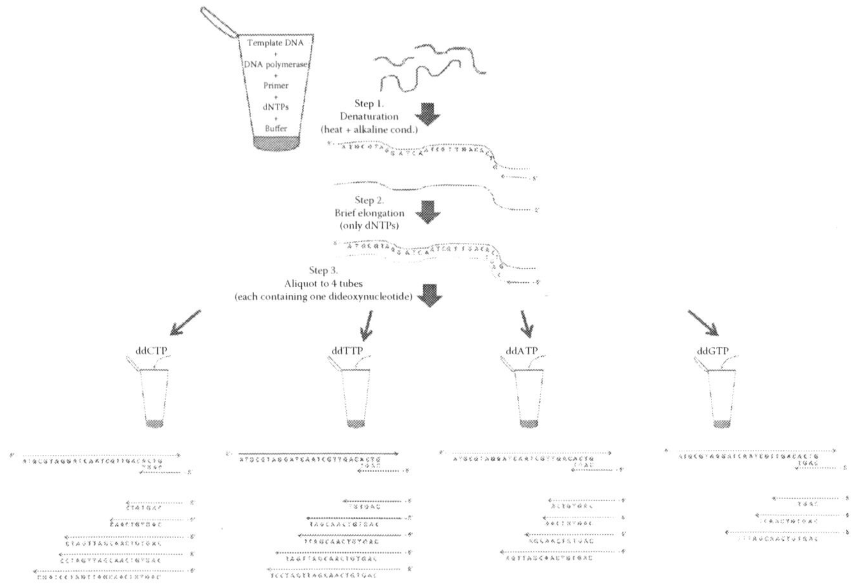

Figure A.4. A diagram of the dideoxy sequencing strategy. Unlabeled deoxynucleotides are used to briefly initiate the sequencing reaction, by DNA polymerase extending the nascent DNA with the help of a sequencing primer and dNTPs after denaturation of the template DNA. Thereafter, the reaction mix is typically aliquoted into four tubes, to each of which one dideoxynucleotide (ddNTP) is added. In each reaction tube, the DNA polymerase will incorporate normal dNTPs, but will also have an equal opportunity of incorporating a ddNTP. For example, in the ddCTP tube, the polymerase will either incorporate a dCTP or a ddCTP with equal probability opposite a "G" on the template strand. The resulting reaction products (which contain a mixture of different sized products in each tube) can either be run on a sequencing gel in classical Sanger dideoxy sequencing (where the nascent DNA will have to be radioactively labeled in the first part of the reaction), or in automated sequencing where the ddNTPs can each be labeled with a different fluorescent label (typically ddCTP is labeled blue, ddTTP is labeled red, ddATP is labeled with green, and ddGTP is labeled with yellow fluorescent dye), run on capillary electrophoresis and analyzed by computer.

SOLiD (Sequencing by Oligonucleotide Ligation and Detection) sequencing relies on ligation, where oligonucleotides of a fixed length are captured on magnetic beads (a library of one fragment per bead), with an adapter sequence. When a universal primer anneals, a library of probes is added to the reaction, allowing the ligation of the probe to the primer, followed by ligation at each cycle (Mardis 2007).

The term **deep sequencing** is used in a different context—it refers to the number of times a sequence is read, or the coverage. *Deep* sequencing is used for reads of seven times or more, whereas *ultra-deep* is now being used for over 100 reads, giving a higher coverage. This coverage is particularly useful if single nucleotide polymorphisms (SNP) are being investigated, since in a single read there are likely to be sequencing errors, whereas when multiple reads are used, it becomes statistically more significant to determine which alterations are simply sequencing errors, and which ones are true SNPs.

Another breakthrough came with the application of microfluidics and microwell technologies to sequencing, making single-cell genomic DNA sequencing as well as RNA sequencing and transcriptomics possible, which was chosen as the "Method of the Year 2013" by *Nature Methods* (Editorial 2014b). This technology has the advantage of making clinical applications or work with rare or a limited number of cells far easier than any other technology.

> Invention is the mother of necessity.
>
> **Thornstein Veblen (1857–1929)**

Appendix B: Protein Techniques

As with the other appendices, it should be noted that the information presented here merely aims to assist the reader with the text and the end-of-chapter questions; there are no laboratory protocols. For more detailed information, the readers are referred to classical biochemistry or molecular biology textbooks (Berg et al. 2002, a.k.a. "The Stryer"; and Nelson and Cox 2008, a.k.a. "Lehninger").

B.1 SDS-PAGE

SDS-PAGE stands for *S*odium *D*odecyl *S*ulfate *P*oly*A*crylamide *G*el *E*lectrophoresis. As with DNA agarose gel electrophoresis, in this process, proteins will be moving through polyacrylamide gels in an electrophoretic field (although nucleic acids can also be run in polyacrylamide gels instead of an agarose gel if precise separation is required). Ammonium persulfate (APS) and tetramethylethylenediamine (TEMED) are commonly used as polymerization stabilizers. SDS is used as a chemical denaturation agent, thereby unfolding the protein, and thus eliminating conformation-related mobility differences. SDS also applies a negative charge to proteins, which is crucial to further eliminate mobility differences in proteins due to charges in the amino acid side chains. In addition to SDS, protein samples are usually also heated, which helps complete unfolding, and DTT (dithiothreitol) and/or 2-mercaptoethanol are used to completely get rid of disulfide linkages or any remaining secondary structures. Therefore, SDS-PAGE can be used to separate proteins purely based on size (charge and conformation effects are eliminated).

To ensure that all proteins are well separated, a two-tier gel system is used—on the top tier there is a low percentage "stacking" gel, typically around 5 to 6% polyacrylamide, and on the bottom tier there is a higher percentage "separating" or "resolving" gel, typically between 9 to 20% polyacrylamide, depending on the size range of the proteins to be separated and studied (higher percentage gels are required for better separation of smaller proteins) (Figure B.1).

Visualization of proteins can be done using specific dyes such as Coomassie Brilliant Blue R-250 (commonly referred to as Coomassie blue) or silver stain.

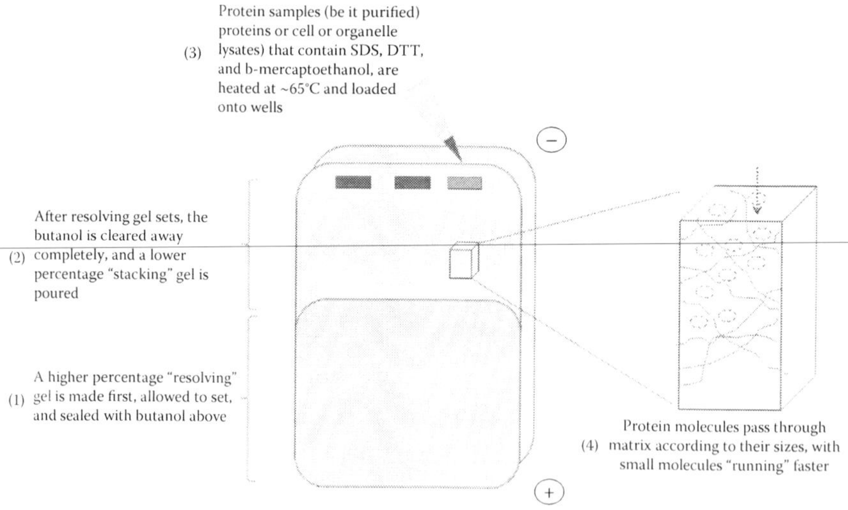

Figure B.1 A basic scheme for SDS-PAGE for the size-based separation of proteins. See the text for details.

B.2 Western Blotting

Western blotting or immunoblotting is a method used to identify specific proteins in a sample, using the antibody–antigen affinity principle (named after the *Southern blot*). Usually, proteins are separated by SDS-PAGE electrophoresis and thereupon transferred to a blotting membrane (which could be nitrocellulose, polyvinylidene fluoride, PVDF, or similar) using again an electrical field for transfer (hence, called *electroblotting*) (Figure B.2). There are wet blots, semidry blots, or dry-blots from different commercial companies, all of which could be used according to the manufacturer's instructions and based on personal preference and experience.

Detection is done using a combination of primary and secondary antibodies, as shown in Chapter 4, Figure 4.7 (also see Figure B.2). For very basic information about antibodies, please refer to Box 4.1 of Chapter 4. After incubation with primary and secondary antibodies, detection is usually done through an enzymatic or biochemical reaction involving the conjugate molecule of the secondary antibody (Figure B.2).

The most common enzyme conjugates used with secondary antibodies are the alkaline phosphatase (AP) and the horseradish peroxidase (HRP), although many more are available. AP is a typical phosphatase enzyme that removes phosphate groups from substrates. The most common AP substrate used in Western blots is the 5-bromo-4-chloro-3-indolyl phosphate (BCIP) and nitroblue tetrazolium (NBT), which are processed by AP to yield an indigoid dye (purple) and an insoluble diformazan precipitate (blue), respectively, in colorimetric reactions. HRP, on the other hand, is a typical peroxidase that can be used both in colorimetric detection and chemiluminescent reaction: when 4-chloro-1-naphthol (4CN) and 3,3′-diaminobenzidine (DAB) are used as HRP substrates in Western blots, the enzyme converts them to an insoluble purple product and an insoluble brown precipitate, respectively. When HRP reacts with luminal, a chemiluminescent substrate,

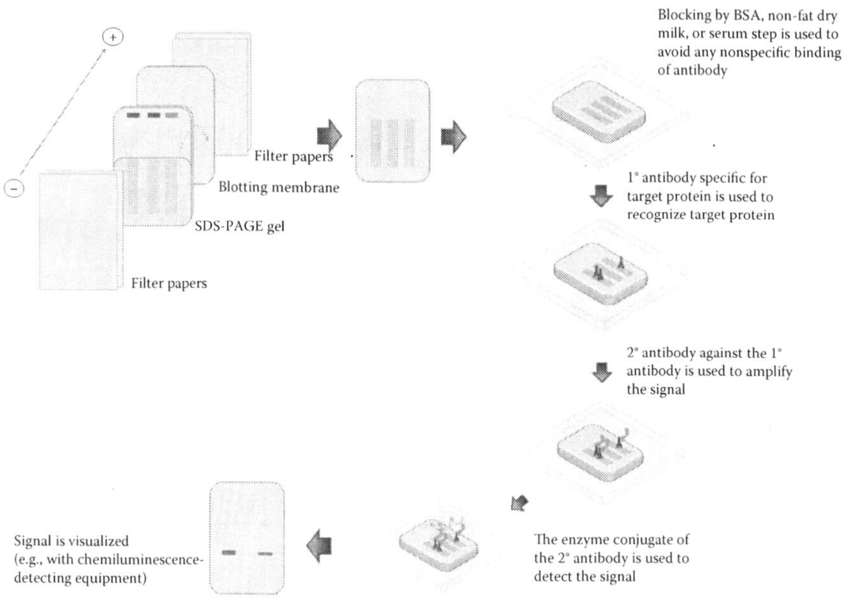

Figure B.2 A basic schematic of the Western blotting principle. Proteins separated by SDS-PAGE (see Figure A2.1) are electroblotted onto an appropriate blotting membrane, after which the *blot* is first *blocked* using BSA, nonfat dry milk or similar, so as to avoid any non-specific binding of primary antibody to random proteins. After washing with a phosphate-buffered, saline-based, or similar wash solution, the blot is immediately incubated with the appropriate primary (1°) antibody that is specific to the target protein (or the tag it is fused to; see Chapter 4). Again, after a wash step, the blot is then incubated with the appropriate secondary (2°) antibody that recognizes the 1° antibody and is conjugated (typically) with an enzyme, which can then be used to detect the signal through electrochemiluminescence or a color reaction.

on the other hand, it will produce light, which can be detected and quantified by chemiluminescence detecting equipment available from various manufacturers. (Chemiluminescent substrates for AP are also available from some companies, however, HRP conjugate has been far more common in this detection method.)

B.3 2D Gel Electrophoresis and Proteomics

Another analytical technique in biochemistry, which readers can find more detailed information about in the aforementioned classical biochemistry books, is 2-dimensional (2D) gel electrophoresis. The two dimensions are required for higher resolution separation of proteins: in the first dimension there is separation of nondenatured proteins (based mainly on charge, called *isoelectric focusing*), and in the second dimension there is a purely size-based separation in the denaturing gel (similar to SDS-PAGE).

The principle is this: in the first dimension, the isoelectric focusing separates proteins based essentially on charge, through a gel that contains a pH gradient (which can be adjusted according to the protein groups to be studied). Proteins move along the gel to either the positive or negative end, until they reach their isoelectric point, at which point the net charge on the protein will be zero, and hence the proteins will no longer have mobility. After the first dimension, the gel (with different proteins settled at their respective isoelectric points) will be *mounted* onto an SDS-PAGE denaturing gel, whereby proteins from the first dimension will now be separated according to size, as described above, in this second dimension (Figure B.3). The incorporation of the first dimension greatly improves resolution, both separating proteins of similar size but different amino acid compositions, but also separating proteins that are post-translationally modified (such as phosphorylated or acetylated proteins) with higher precision than standard SDS-PAGE gels.

The proteins are then commonly detected by silver staining, or Coomassie staining described above, appearing as *spots* on the gel, which are then analyzed with the help of specialized computer programs. The spots of interest can then be excised from the gel, and analyzed by mass spectrometry (Figure B.3).

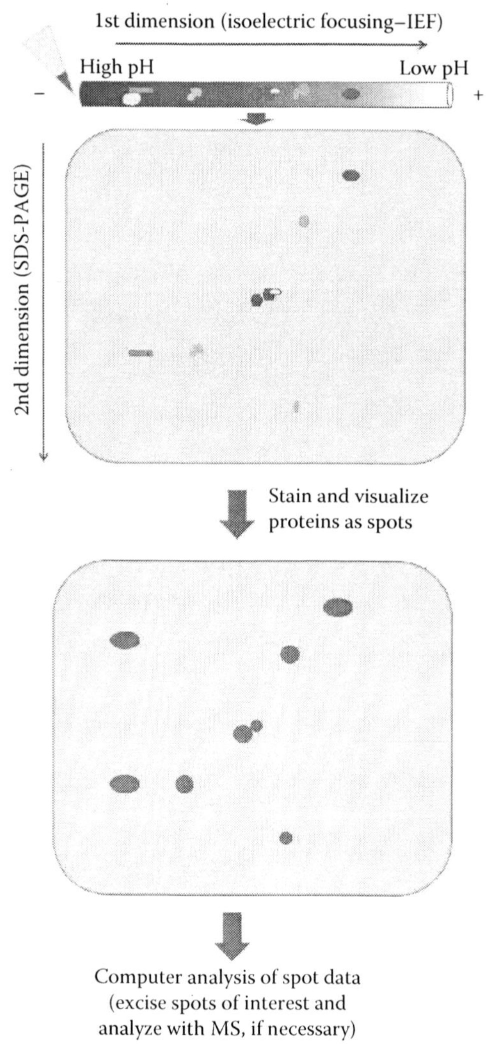

Figure B.3 A diagram showing the basic principle of 2D gel electrophoresis. See the text for details.

Proteomics is the large-scale and global analysis of (almost) all proteins (and every post-translational variant thereof) in a cell or tissue. A 2D gel electrophoresis–mass spectrometry combination is perhaps the most commonly used approach to study whole proteome profiles, although protein microarrays can also be used. However, since proteins have to be detected with antibodies (with current technologies), and since both proteins and antibodies can be denatured quite easily, affecting their interaction, protein arrays or protein chips are more difficult to optimize, and can thus not be routinely employed. Differential gel electrophoresis (DIGE), which uses Cy2, Cy3, and/or Cy5 labeling of different protein samples, allows for higher sensitivity of the fluorophore-labeled proteins. Other biochemical techniques such as liquid chromatography followed by tandem mass spectrometry (LC/MS/MS) are also used in proteomics.

Mass spectrometry-based targeted proteomics, which can detect specific proteins of interest with *high sensitivity and reproducibility*, has in principle been around for quite some time, and yet because of its widespread use and the quantitative information recently made possible, it was chosen the "Method of the Year 2012" (Editorial 2013). *Tissue proteomics*, from formalin-fixed and paraffin-embedded (FFPE) tissues, analyzed by matrix-assisted laser desorption/ionization (MALDI) MS imaging, has become popular, as imaging of hundreds of compounds on a tissue section is possible, linking histology with molecular biology and proteomics (Lonquespee et al. 2014).

B.4 Immunofluorescence and Immunohistochemistry

Both these techniques rely on the detection of proteins in cells or tissues using antibodies (*immuno-*), either by fluorescent or enzyme conjugate on the secondary antibody, that is, either by fluorescent or light microscopy. In immunohistochemistry, *-histo* refers to the staining of tissues; if dispersed cells (devoid of any extracellular matrix) are stained, then the technique is also called *immunocytochemistry (-cyto-* refers to staining of individual cells).

The basis of both immunofluorescence and immunohistochemistry is in principle the same, using primary antibodies as probes to recognize and identify target proteins, and secondary antibodies (in indirect fluorescence) against the primary antibody to amplify the signal (Figure B.4). The only difference is the cells or tissues being used, which will slightly change the protocol (depending on whether the extracellular matrix is around or not; for examples, please refer to detailed protocols).

In essence, the basic logic is the same as in Western blots—one is trying to detect the expression of a protein. However, since cells are used, information about the subcellular localization of proteins can also be obtained in addition to the level of expression. (Note: The technique is not limited to proteins—any molecule against which a specific antibody is available can be studied). The first step is the *fixation* of cells or tissues onto the microscope slide, followed by permeabilization. This step is essential, not only because cells must remain on the slide at all times and not get washed away, but also so that antibodies against target molecules can penetrate through the cell membrane (and cell wall, if yeast, fungi, or plants are studied). Following fixation/permeabilization, a *blocking* step is necessary, for the same reasons as in Western blotting (i.e., to prevent nonspecific binding of antibodies).

Once slide preparations are blocked and excess reagents are washed away, the slides are incubated with primary antibodies against (and specific for) the target protein. (Direct immunofluorescence, in fact, uses conjugated primary antibodies and directly prepares slides for detection; however, this is not employed frequently.) Following the primary antibody incubation, the slides are briefly washed, and incubated with secondary antibodies against the primary. This secondary antibody contains either a fluorescent conjugate, such as fluorescein isothiocyanate (FITC), which emits a green fluorescence when excited, or an enzyme conjugate, such as HRP, which can be used for colorimetric detection in light microscopy.

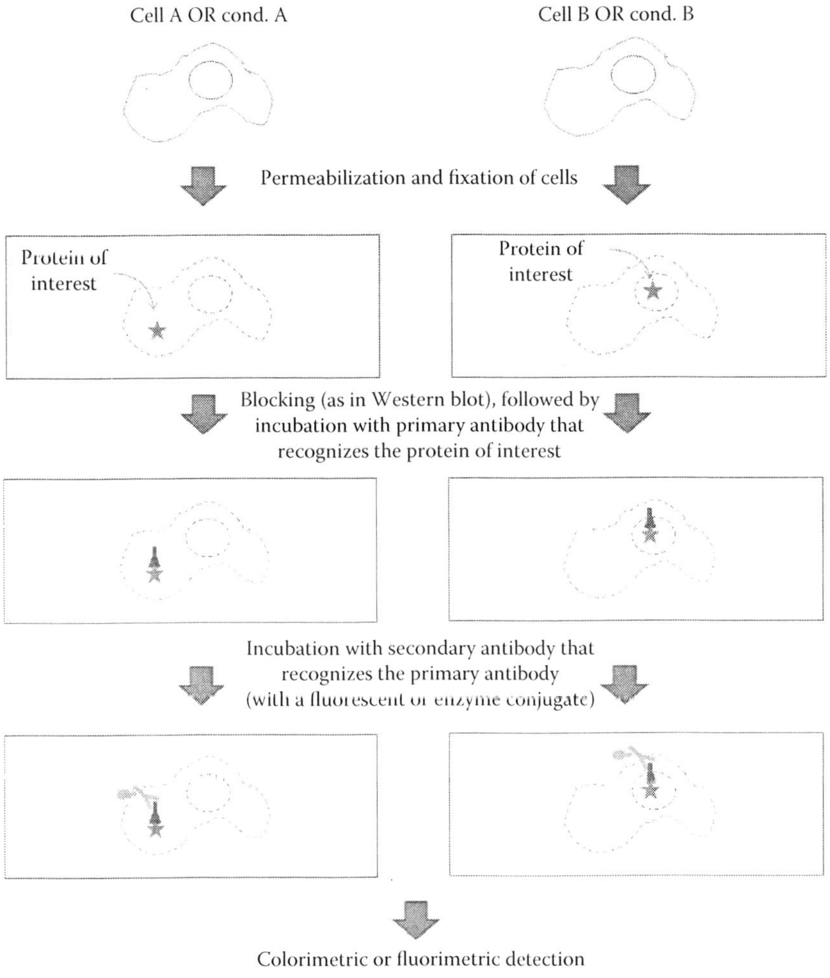

Figure B.4 A brief overview of the steps of immunofluorescence and immunohistochemistry. Two samples are shown here: either a protein's amount and localization can be studied in cell A versus cell B, or the level of expression and localization of a protein can be studied under two (or more) different conditions (cond. A and cond. B). See the text for details.

Appendix C: Supplement Information for End-of-Chapter Questions

C.1 Nucleotides and Nucleic Acids

The structure and biochemistry of nucleotides and nucleic acids is beyond the scope of this book. However, as a quick reference or guide, the following schematics are provided in this technical data appendix (Figure C.1). If more detailed biochemical information is required, the readers are referred to classical biochemistry textbooks.

C.2 Genetic Coding Tables

The genetic code, a term that was coined by Francis Crick in his revolutionizing paper "General Nature of the Genetic Code for Proteins" (Crick et al. 1961) and later elucidated by the pioneering work of Nirenberg and friends (Leder and Nirenberg 1964; Matthaei et al. 1962; and many others), has a critical importance in determining the protein sequence, hence, the function. Three of the 64 triplicate nucleotide codons code for stop for translation, while the remaining 61 codons of the so-called Universal Genetic Code code for the 20 amino acids (Table C.1). Having said that, alternative genetic code tables (25 of them in

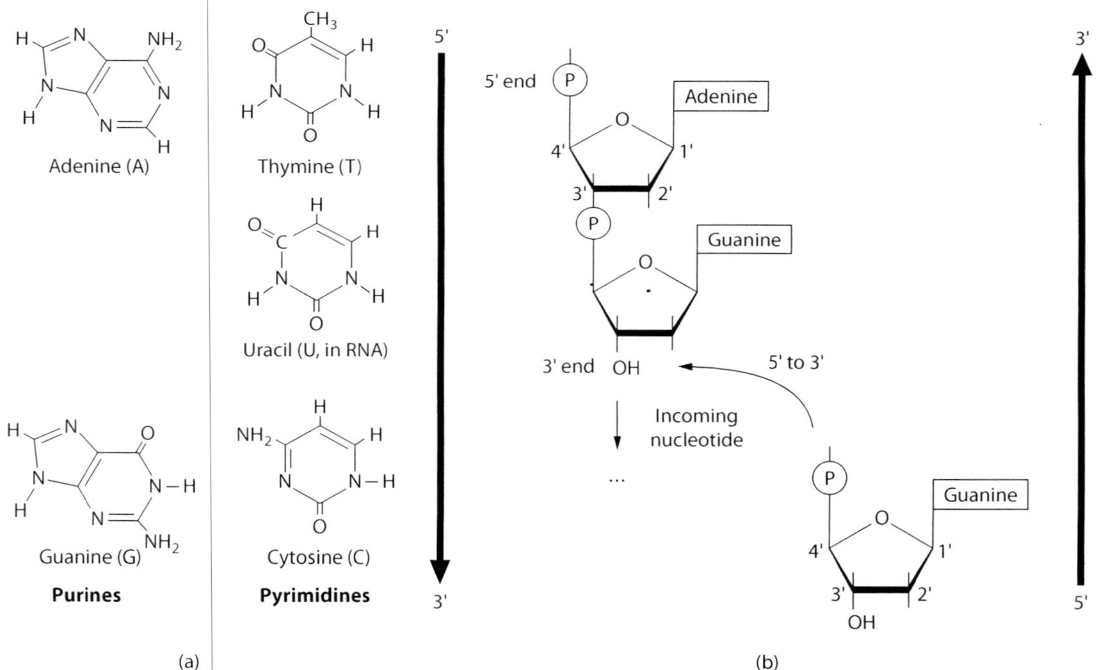

Figure C.1 **A diagram of nucleotides and nucleic acid polymerization. (a) The common nucleotides found in DNA and RNA molecules, categorized as purines (left) and pyrimidines (right). (b) The principle of how a nucleic acid is polymerized through a sugar–phosphate bond formation between the 5′- phosphate group of an incoming nucleotide and the 3′- hydroxyl group of the nascent nucleic acid chain.**

Table C.1 The Universal (or Genetic) Code Table

	U	C	A	G
U	UUU Phe (F) UUC Phe (F) UUA Leu (L) UUG Leu (L)	UCU Ser (S) UCC Ser (S) UCA Ser (S) UCG Ser (S)	UAU Tyr (Y) UAC Tyr (Y) **UAA Stop** **UAG Stop**	UGU Cys (C) UGC Cys (C) **UGA Stop** UGG Trp (W)
C	CUU Leu (L) CUC Leu (L) CUA Leu (L) CUG Leu (L)	CCU Pro (P) CCC Pro (P) CCA Pro (P) CCG Pro (P)	CAU His (H) CAC His (H) CAA Gln (Q) CAG Gln (Q)	CGU Arg (R) CGC Arg (R) CGA Arg (R) CGG Arg (R)
A	AUU Ile (I) AUC Ile (I) AUA Ile (I) AUG Met (M)	ACU Thr (T) ACC Thr (T) ACA Thr (T) ACG Thr (T)	AAU Asn (N) AAC Asn (N) AAA Lys (K) AAG Lys (K)	AGU Ser (S) AGC Ser (S) AGA Arg (R) AGG Arg (R)
G	GUU Val (V) GUC Val (V) GUA Val (V) GUG Val (V)	GCU Ala (A) GCC Ala (A) GCA Ala (A) GCG Ala (A)	GAU Asp (D) GAC Asp (D) GAA Glu (E) GAG Glu (E)	GGU Gly (G) GGC Gly (G) GGA Gly (G) GGG Gly (G)

Note: The triplate codon and the three-letter and single-letter amino acid codes in parentheses are written successively.

the July 2014 update of the NCBI Taxonomy database at http://www.ncbi.nlm.nih.gov/Taxonomy/Utils/wprintgc.cgi?mode=c) also exist, which are used either by various organelles such as mitochondria and plastids, or by various organisms such as the ciliates (for a review, refer to Knight et al. 2001).

The genetic codes are degenerate, in the sense that some amino acids are coded by multiple codons in all coding tables, some codons only differ in the third position (for example, CCU, CCC, CCG, and CCA all code for proline) (see Table C.1). This property can be used to mutagenize coding sequences in such a way that, for example, a restriction enzyme recognition motif is introduced to the sequence

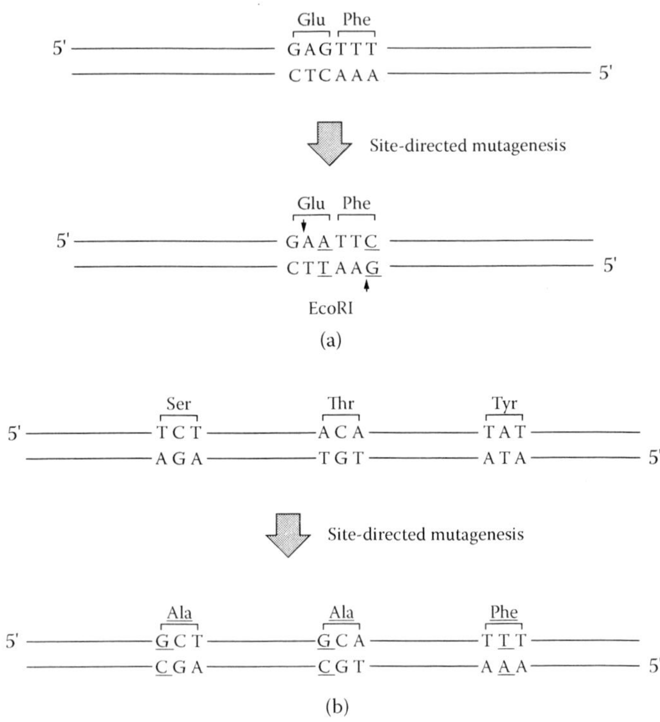

Figure C.2 A schematic of hypothetical site-directed mutagenesis examples. (a) Mutagenesis of a coding sequence that incorporates an EcoRI recognition motif without affecting the amino acid sequence coded by that DNA. (b) Mutagenesis of a coding sequence that changes specific amino acids into nonphosphorylatable alanine or phenylalanine. Mutagenized nucleotides are underlined; amino acids encoded by the indicated triplicate codon is given above the sequence, and the restriction motif, if any, is given below the sequence. Small arrows indicate the cut sites in (a).

without affecting the protein coded by that sequence (Figure C.2a). Alternatively, a phosphorylation site, for example, can be *knocked out* by simply making that amino acid *unphosphorylatable*—typically a serine or threonine, which contain an –OH (hydroxyl) group to which the phosphate group is added, is mutated into an alanine, or a tyrosine is converted to a phenylalanine (Figure C.2b) (see Figure 4.3 in Chapter 4 for amino acid structures).

C.3 Amino Acids

The detailed biochemistry of amino acids, too, is beyond the scope of this book, however, the following diagram is given only as a reference to the main text and as an aid to questions at the end of each chapter (Figure C.3). If further and more detailed biochemical information is required, the readers are referred to classical biochemistry textbooks.

C.4 Calculations Regarding Nucleic Acids, Amino Acids, and Proteins

When designing experiments, one needs not only to calculate the concentrations of solutions to be used in the cloning protocols, but also to correlate the coding DNA sequence to proteins synthesized by that sequence, do the molar conversions or calculate the sizes of the fusion proteins resulting from the fusion clonings, and so on. The following information is considered a handy supplement to many of the questions present in each chapter of this book.

C.4.1 Spectrophotometric Measurements

The following calculations are commonly employed to estimate the nucleic acid concentration or amount in your solution through spectrophotometric measurement. (Note: "A" stands for *Absorption units*.)

1 A260 unit of double-stranded DNA = 50 ug/ml
1 A260 unit of single-stranded DNA = 33 ug/ml
1 A260 unit of single-stranded RNA = 40 ug/ml

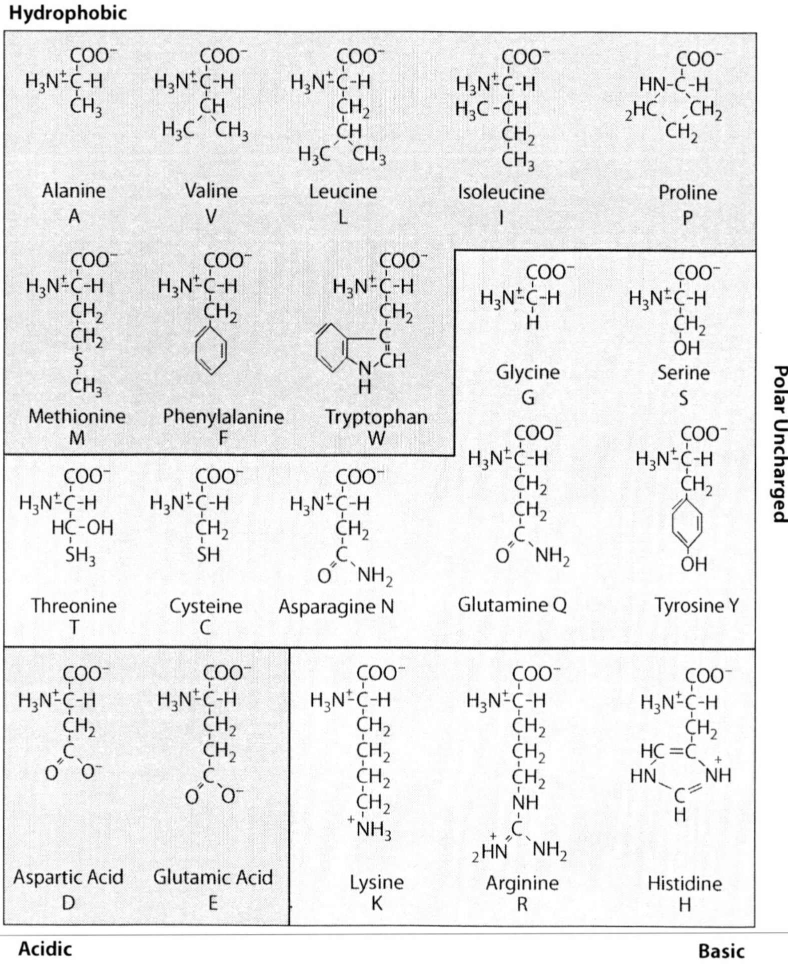

Figure C.3 Properties of amino acids. Hydrophobic, polar uncharged, acidic, and basic amino acids are grouped, along with their structures and single letter designations. (Modified from Docstoc, http://www.docstoc.com/docs/28866577/Notes_-Amino-Acids-and-Proteins.)

C.4.2 Average Molecular Weights

If you need a rough estimate of the molecular weight of your nucleic acid or protein, below are the average estimates:

Average molecular weight of a nucleotide = 330 Da (for DNA)
Average molecular weight of a nucleotide = 340 Da (for RNA)
Average molecular weight of an amino acid = 110 Da

C.4.3 Conversions of Molecular Weights for Protein and DNA

If you need a rough estimate of the size of a protein encoded by a certain region of DNA (of known base pairs), for example, you could use the following estimates:

1 kb of ds DNA = 6.6. × 10^5 Da	333 amino acids = 37 kDa protein
1 kb of ss DNA = 3.3. × 10^5 Da	100 pmol of 1 kDa protein = 100 ng
1 kb of ss RNA = 3.4 × 10^5 Da	1 pmol of 1 Kb DNA = 0.66 μg

(**Note:** *ds* stands for *double stranded*, *ss* stands for *single stranded*, *Da* stands for *Dalton*, *kDa* stands for *kilo Dalton*, and *Kb* stands for *kilo base pairs*.)

C.5 Compatible Overhangs

When cloning, sometimes the desired restriction motif may not be present, or it may not be possible to use the desired restriction motif for one reason or another. Compatible overhangs may then be useful, since they hybridize or anneal to each other as if the DNA sequences were digested by the same restriction enzyme (covered in Chapter 2, Section 2.1).

Figure C.4 shows some of the commonly used restriction enzymes, with overhangs generated indicated above and

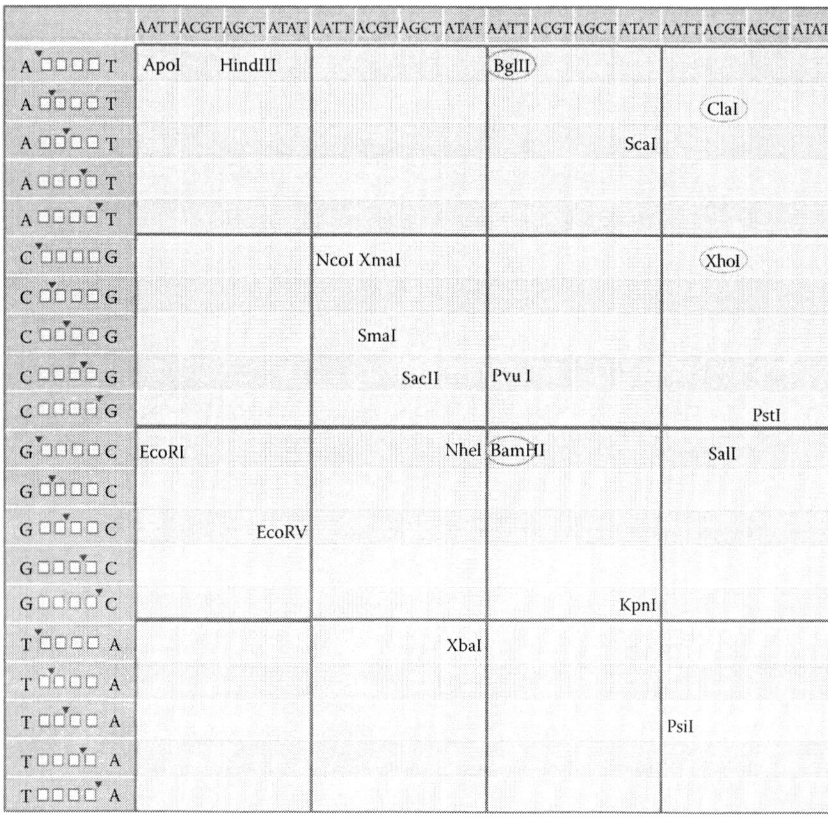

Figure C.4 A table showing recognition motifs and cut sites of some common restriction enzymes. On the left, the cut sites are shown with small solid triangles. The sequence that corresponds to the four squares on the left is found at the top of each column. Enzymes that are found within the same column, at the same row position, generate compatible overhangs (e.g., BglII and BamHI, circles, both generate 5′-AATT overhangs); whereas enzymes found in different row positions, even if they are within the same column, do not (e.g., ClaI and XhoI, circles, generate overlapping but different overhangs, which would interfere with ligation).

restriction motif sequences and cut sites indicated on the left. Restriction enzymes that generate compatible overhangs usually appear in the same column and position (see legend for details).

C.6 Genotypes of Frequently Used Laboratory Strains of Bacteria and Yeast

There are quite a number of bacteria or yeast strains, and their derivatives, which are used for slightly different purposes, such as cloning, mutagenesis, or library generation. Here we have only listed a few of the more common ones. All organisms are assumed to be wild-type, except for the mutations listed, and bacterial strains are λ^- unless mentioned otherwise (Tables C.2 and C.3).

The *dam* and *dcm* loci in bacterial genomes code for DNA methyltransferases that result in methylation of either an adenine or a cytosine, respectively, using an S-adenosylmethionine as a methyl donor (Table C.4).

Methylation is an important parameter to consider when designing restriction or mutagenesis assays, since methylation of DNA within the recognition motif can block or significantly interfere with the catalytic activity of restriction enzymes. For example, restriction enzyme recognition motifs that are blocked by Dam or Dcm methylases can be unmethylated by cloning the plasmid DNA into a *dam*⁻, *dcm*⁻ strain. (One needs to check methylation sensitivities of restriction enzymes before designing an experiment.)

C.7 Methylation Sensitivities of Common Restriction Enzymes

Table C.5 shows methylation sensitivities of some common restriction enzymes.

Table C.2 Genotypes of Some Common Laboratory Strains of Bacteria and Yeast

Common Bacterial Strains and Genotypes	
Strain	*Genotype*
BL21(DE3)	F–, *ompT*, *hsd*S_B(r_B–, m_B–), *dcm*, *gal*, λ(DE3)
BL21(DE3) pLysS	F–, *ompT*, *hsd*S_B(r_B–, m_B–), *dcm*, *gal*, λ(DE3), **pLysS (Cm^r)**
DH5α™	φ80d*lac*ZΔM15, *rec*A1, *end*A1, *gyr*AB, *thi*-1, *hsd*R17(r_K–, m_K+), *supE*44, *relA*1, *deo*R, Δ(*lacZYA-arg*F) U169, *pho*A
HB101 (4)	*thi*-1, *hsdS*20(r_B–, m_B–), *sup*E44, *rec*A13, *ara*-14, *leuB*6, *proA*2, *lacY*1, *gal*K2, *rpsL*20(str^r), *xyl*-5, *mtl*-1
JM109 (5)	*endA*1, *recA*1, *gyr*A96, *thi*-1, *hsdR*17(r_K–, m_K+), *relA*1, *supE*44, Δ(*lac-proAB*), [F′, *traD*36, *proAB*, *lac*I^qZΔM15]
SURE®	e14–, (*mcr*A–), Δ(*mcr*CB-*hsd*SMR-*mrr*)171, *end*A1, *sup*E44, *thi*-1, *gyr*A96, *rel*A1, *lac*, *rec*B, *rec*J, *sbc*C, *umu*C::Tn5(*kan*^{r}), *uvr*C, [F′, *proAB*, *lac*I^qZΔM15::Tn10(tet^r)]
TOP10	F–, *mcr*A, Δ(*mrr-hsd*RMS-*mcr*BC), φ80*lac*ZΔM15, Δ*lac*X74, *deo*R, *rec*A1, *ara*D139, Δ(*ara*, *leu*)7697, *gal*U, *gal*K, *rps*L(strr), *end*A1, *nup*G
XL1-Blue	*rec*A1, *end*A1, *gyr*A96, *thi*-1, *hsd*R17(r_K–, m_K+), *sup*E44, *rel*A1, *lac*, [F′, *proAB*, *lac*I^qZΔM15::Tn10(tet^r)]
XL10-Gold®	Tet^r, Δ(*mcr*A)183, Δ(*mcr*CB-*hsd*SMR-*mrr*)173, *end*A1, *sup*E44, *thi*-1, *rec*A1, *gyr*A96, *rel*A1, *lac* Hte, [F′, *proAB*, *lac*I^qZΔM15,Tn10(tet^r) Amy Cam^r][a]
JK9-3da	*MAT*a *leu2-3,112 ura3-52 rme1 trp1 his4*
S288C	*MATα SUC2 gal2 mal2 mel flo1 flo8-1 hap1 ho bio1 bio6*
SK1	*MATa/α HO gal2 cupS can1R BIO*
W303	*MATa/MATα {leu2-3,112 trp1-1 can1-100 ura3-1 ade2-1 his3-11,15} [phi+]*

Note: The mutant alleles in Table C.2 are summarized in Table C.3.

Table C.3 Brief Descriptions of Mutant Alleles Listed in Table C.2

Mutation	*Description*
araD	Mutation of the enzyme L-ribulose phosphate 4-epimerase; blocks arabinose catabolism
dcm	Mutation of the enzyme DNA cytosine methylase
deoR	Mutation of the regulatory gene, which results in constitutive expression of genes for deoxyribose synthesis, thus allowing efficient propagation of plasmids
endA1	Mutation in DNA-specific endonuclease I, which improves the quality of DNA isolation
galK	Mutation in galactokinase enzyme, which blocks galactose catabolism
gyrA96	Mutation in DNA gyrase that confers resistance to nalidixic acid, an inhibitor of bacterial growth
hsdR	Mutation of the host DNA restriction and methylation system, which allows cloning without cleavage of transformed DNA by endogenous restriction endonucleases
hsdS20	Mutation of specificity determinant for host DNA restriction and methylation system. This mutation allows cloning without cleavage of transformed DNA by endogenous restriction endonucleases
lacY	Mutation of galactoside permease, which blocks lactose utilization
lacZΔM15	Partial deletion of β-D-galactosidase gene, which allows complementation of β-galactosidase activity in blue-white screening when plated on X-gal
leuB	Mutation of the β-isopropylmalate dehydrogenase gene, which necessitates presence of leucine on minimal media for growth
LysS	pLysS plasmid integrated into the genome, which will confer additional antibiotic resistance (chloramphenicol or tetracycline) and produce T7 lysozyme that inhibits T7 RNA polymerase and lower background transcription from T7 RNA polymerase promoter

continued

Table C.3 (continued) Brief Descriptions of Mutant Alleles Listed in Table C.2

Mutation	*Description*
mcrA	Mutation of the methylcytosine restriction system, which will block restriction of DNA methylated at the sequence 5′-G^{m}CGC-3′
mtl	Mutation in the mannitol metabolism, blocking catabolism of mannitol
ompT	Mutation of protease VII, an outer membrane protein, which reduces proteolysis of expressed proteins
proA	Mutation in γ-glutamyl phosphate reductase, which will result in excretion of proline on minimal media, and confer resistance to proline analogs
recA	Mutations in recombination and repair
relA	Mutation in ppGpp synthetase I enzyme, allowing RNA synthesis in the absence of protein synthesis
rpsL	30S ribosome S12 subunit mutation, conferring resistance to streptomycin
sbcB	Mutation in Exonuclease I, allowing general recombination in recBC mutant strains
supE	Suppressor mutant that suppresses amber (UAG) mutations
thi-1	Mutations in thiamine metabolism, which necessitates thiamine supplement for growth
umuC	Mutation in umuC UV mutagenesis and repair protein, a component of the SOS response
uvrC	Mutation in uvrC UV mutagenesis and repair protein, a component of the nucleotide excision repair mechanism
xyl-5	Mutation in xylose metabolism, which blocks catabolism of xylose

Table C.4 Recognition and Methylation Sites of Common DNA Methylases

Methylase	*Site Recognized (Nucleotide Methylated Is Indicated in Boldface)*
Dam methylase	N^6 position of the adenine in the sequence GATC
Dcm methylase	C^5 position of cytosine in the sequences CCAGG and CCTGG
EcoKI methylase	Adenine in the sequences AAC(N^6A)GTGC and GCAC(N^6A)GTT

Table C.5 Methylation Sensitivities of Some Common Restriction Enzymes

Enzyme	*Recognition Motif*	*dam*	*dcm*
BamHI	GGATCC	i	I
BglII	AGATCT	i	I
ClaI	ATCGAT	s(ol)	i
EcoRI	GAATTC	i	I
HindIII	AAGCTT	i	I
NdeII	GATC	s	i
NheI	GCTAGC	i	i
SacII	CCGCGG	i	I
SalI	GTCGAC	i	I
ScaI	AGTACT	i	I
SmaI	CCCGGG	i	I
XbaI	TCTAGA	s(ol)	I
XhoI	CTCGAG	i	I
XmaI	CCCGGG	i	i

Note: s, sensitive to methylation; *i*, insensitive to methylation; *s(ol)*, sensitive only when restriction site overlaps the methylation sequence.

C.8 Primary and Secondary Antibody Examples

Primary and secondary antibodies can be used in a wide range of different protein detection, purification, or interaction assays, as described in Chapters 4 and 6. This section merely provides a sample selection of antibodies so as to supplement the chapters as well as help with the end-of-chapter questions.

Table C.6 summarizes the basic features of some common primary antibodies that could be used when answering some of the end-of-chapter questions.

Table C.7 summarizes the basic features of some common secondary antibodies that could be used when answering some of the end-of-chapter questions.

Table C.6 Primary Antibodies

Antibody Name	*Source*	*Applications*	*Reactivity*
α-GST	Rabbit IgG	WB, IP, IF	m
α-GST	Mouse IgG	WB	m,r,h
α-GST	Mouse IgG	WB, IP, IF	h
α-HA	Mouse IgG	WB, IP	N/A
α-HA	Rat IgG	WB, IP, IF, FCM	N/A
α-Myc	Rabbit IgG	WB, IP, IF	N/A
α-Myc	Mouse IgG	WB, IP	N/A
α-Flag	Mouse IgG	WB, IP, FCM	N/A
α-Flag	Rabbit IgG	WB, IP	N/A
α-actin	Mouse IgG	WB, IP, IHC	m,r

Note: α-, anti-; WB, Western blot; IP, immunoprecipitation; IF, immunofluorescence; FCM, fluorescent confocal microscopy; IHC, immunohistochemistry; m, mouse; r, rat; h, human; N/A, not applicable.

Table C.7 Secondary Antibodies

Antibody Name	*HRP Conjugate*	*FITC*	*Texas Red*
Chicken anti-mouse IgG	+	+	+
Donkey anti-mouse IgG	+	+	+
Goat anti-mouse IgG	+	+	+
Goat anti-rabbit IgG	+	+	+
Goat anti-rat IgG	+	+	–
Chicken anti-rat IgG	+	–	+

Note: HRP, horseradish peroxidase; FITC, fluorescein isothiocyanate.

C.9 Fluorophores

Fluorophores are described and categorized according to their absorption and emission spectra and maxima. (For a brief description, see Chapter 6, Section 6.4 and Figure 6.7.)

Table C.8 is a quick reference table for some common fluorophores and their maximum excitation and emission wavelengths.

Table C.8 Table Summarizing Common Fluorophores and Their Excitation/Emission Maxima

Fluorophore	*Excitation (nm)*	*Emission (nm)*
DAPI	345	458
Hoechst 33258	365	480
GFP	395/489	509
FITC	494	518
CY3	548	562
CY5	650	670–700
Texas Red	589	615
dTomato	590	620

Note: nm, nanometers.

C.10 Radioisotopes in Biology

The number of protons for each atom is constant; however, the number of neutrons can vary, which are termed **isotopes**. Some are just **heavy isotopes**, for example ^{1}H and ^{2}H are both hydrogen atoms, with a single proton in their nuclei. Yet ^{2}H (also called *deuterium*) has one extra neutron than ^{1}H, only making it heavier without affecting the stability of the nucleus.

Radioactive isotopes, or **radioisotopes**, however, have *unstable nuclei* and as such prefer to disintegrate into more stable nuclei by emitting subatomic particles, such as an alpha particle (consisting of two protons and two neutrons) or a beta particle (consisting of an electron or a negatively charged nuclear particle), and gamma rays (electromagnetic energy waves). ^{3}H, for example, is a radioactive isotope of hydrogen (also called *tritium*). Chemically, however, ^{1}H, ^{2}H, and ^{3}H are all hydrogen atoms and incorporation of either species will not affect any chemical or physiological properties of the molecule, however, ^{2}H and ^{3}H will simply *tag* the molecule in question.

The **half-life** of a radioisotope is defined as the period required for half of the original population of radioactive particles to decay and produce the stable daughter product, and the half-life is specific to each radioisotope species. In cell biology, radioisotopes of common atoms, that is, H, C, S, and P, are routinely used to tag or label various biological macromolecules. The half-life of these molecules can be found in Table C.9. The SI (International System of Units) unit for ionizing radiation is the Becquerel (Bq), after the scientist Henri Becquerel who received the Nobel Prize along with Marie Sklodowska-Curie and Pierre Curie, although researchers commonly use the Curie (Ci) units as a courtesy to Marie and Pierre Curie.

Table C.9 Table Summarizing Common Radioisotopes, Their Half-Lives, and the Type of Particle Emitted

Radioisotope	*Half-Life*	*Particle Emitted*
^{32}P	14.3 days	β
^{33}P	25 days	β
^{14}C	5730 years	β
^{3}H	12.3 years	β
^{35}S	87.4 days	β

1 **Bq** has come to define 1 disintegration per second:

$$1 \text{ Bq} = 1 \text{ s}^{-1}$$

But it can be converted to **Curies** according to the following equation:

$$1 \text{ Bq} = 2.7 \times 10^{-11} \text{ Ci}$$

Energy emitted during radioactive processes are measured in **joules** (J), whereas the **absorbed dose** of radiation (by, for instance, the human body) is defined in terms of **Gray** (Gy), where

$$1 \text{ Gy} = 1 \text{ joule/kg} = 100 \text{ rads}$$

The radiation dose equivalent also takes into account the damage to living tissue, which can vary among species of radioisotopes, and is defined by **Sievert (Sv)**:

For X-rays, gamma rays, or beta rays	1 Sv = 1 Gy
For neutrons	1 Sv ~ = 10 Sv
For alpha particles	1 Sv ~ = 20 Sv

Table C.10 Common Radiolabeled Molecules Used in Molecular Assays

Radioisotope	*Activity*
Adenosine 5′-[α-^{32}P]triphosphate	30 mCi/mmol 74 MBq/ml, 2 mCi/ml
Adenosine 5′-[γ-^{32}P]triphosphate	6000 mCi/mmol 74 MBq/ml, 2 mCi/ml
L-[^{35}S]Cysteine	100–150 mCi/mmol 555 MBq/ml, 15 mCi/ml
Deoxy[2,8-^{3}H]adenosine 5′-triphosphate	5–25 Ci/mmol 37 MBq/ml, 1 mCi/ml
L-[*methyl*-^{3}H]Methionine	70–85 Ci/mmol 185 MBq/ml, 5 mCi/ml
L-[^{35}S]Methionine (*in vitro* translation grade)	> 1000 Ci/mmol 555 MBq/ml, 15 mCi/ml
L-[^{35}S]Methionine (*in vivo* cell labeling grade)	> 1000 Ci/mmol 370 MBq/ml, 10 mCi/ml

Note: α-, β-, and γ- refer to the specific functional group in the molecule that contains the indicated radioisotope.

The presence of radioisotopes can be detected by a number of methods, such as autoradiography or scintillation counters.

Table C.10 contains some common radioisotope-containing molecules used in molecular assays and hypothetical concentrations and activities for end-of-chapter questions.

Appendix D: Answers to End-of-Chapter Questions

Please note that there are fewer sample answers in the last chapters, since a multitude of answers are possible using different combinations of "Lego blocks." Science is limited only by your imagination.

D.1 Chapter 2 Answers

D1.

GAATT AATTC for XmnI
G AATTC for Tsp509I
GAAT TAATTC for AseI

D2.

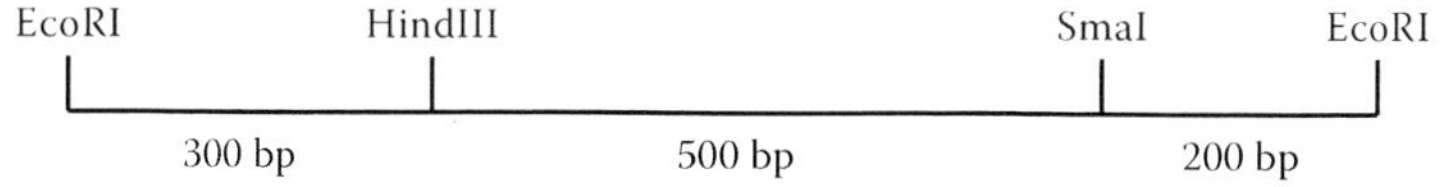

The figure shows the restriction map of the EcoRI-cut DNA fragment.

D4. From the NCBI nucleotide database, one can search for this sequence by typing in the gene name. The accession number is NM_002467. You can then cut the coding sequence (corresponding to the region between ATG and stop codons, as indicated in the descriptions of the NCBI entry) from the NCBI database and paste it into Workbench using Add New Sequence. (You need to choose Session Tool and then Nucleotide Tool, then select Add New Sequence and press Run.) Afterward you would have to choose the user-entered sequence from the sequence list, and the TACG tool from the nucleotide tool menu, then press Run. In the parameters page for TACG, you need to define the criteria mentioned in the question—namely, 6-base cutters and Reading Frame 1. (You have already entered the coding sequence in the correct reading frame—no need to check all six reading frames.)

D7. There are different ways to solve this problem. We will only present some of the options, however, what one needs to pay attention to is the reading frame of both the tag and the inserted protein, reflected also in the restriction sites on both the vector and the insert.

If one chooses to directly use the EcoRI motif present in both the vector and the insert and carry out the cloning, this would cause a shift in the reading frame of the inserted protein sequence, as shown in the figure below, which shows (a) a schematic depiction of one possible solution (b) that creates a frameshift in the insert.

(The right-hand side in this case does not cause any further problem, since the stop codon will terminate translation after the inserted protein; therefore either BglII or XhoI can be chosen on that side.)

Therefore, if you choose to use the EcoRI site close to the start codon, you will need to modify the insert so as to restore the reading frame with respect to that of the tag. One of the convenient ways to do this is through the use of PCR, where you design your primers to contain additional bases, which will restore the frame (in this example, two additional bases need to be added). The figure shows a schematic depiction of an alternative solution that uses PCR-based engineering (a) to generate an insert that is in frame with the HA tag (b).

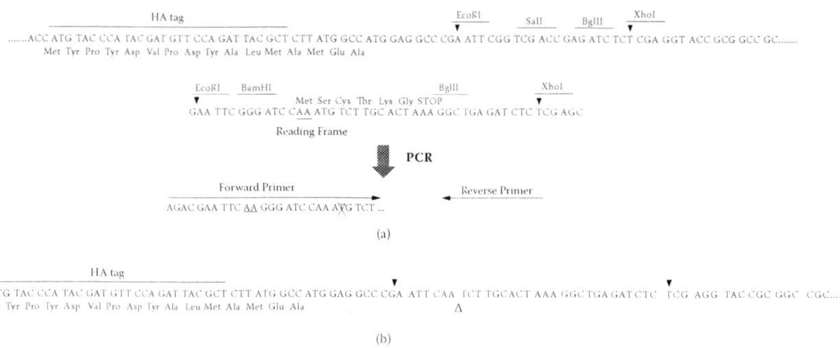

The addition of those two nucleotides (AA, underlined) restores the reading frame (ATG of the insert can also be deleted, if required) parallel to the reading frame of the HA tag. The end of the gene is now *out of frame* with the vector (ligation of the XhoI ends); however, since the translation will terminate at the stop codon of the insert, the "frame" after the TGA is irrelevant for protein synthesis (and again for the same reason, either BglII or XhoI can be used at this end).

Alternatively, you can use another restriction motif which is already in frame with the tag in the vector—the BamHI site in the insert and the BglII site of the vector will produce compatible ends that can be used for cloning, and the frames of these two sites are the same. Therefore, the insert can be digested from the BamHI site on one end, and BglII site on the other end, while the vector gets digested at the BglII site, and the ligation will produce the following product. The figure shows a schematic depiction of another alternative solution that uses a different combination of restriction enzymes (a) that results in the destruction of the BamHI site on the vector (b). In the cloning product it must be noted that once the BamHI and BglII sites are ligated, they will no longer be recognized by either restriction enzyme (since you lose the palindrome). Therefore, for screening a restriction site to the left of the BglII motif one vector must be chosen (either EcoRI or Sal I can be used, as long as these sequences are not present within the insert and produce diagnostic fragments).

D12. (a) Since the sizes of the insert and vector are different (the vector is 10 times the size of the insert), equal weights of the insert and vector mean that for the n molecules of the vector, there are 10n molecules of the insert. See below for a schematic depiction of how the size of a molecule and the length of a molecule (in terms of bp) are related, which in turn are correlated to the number of molecules.

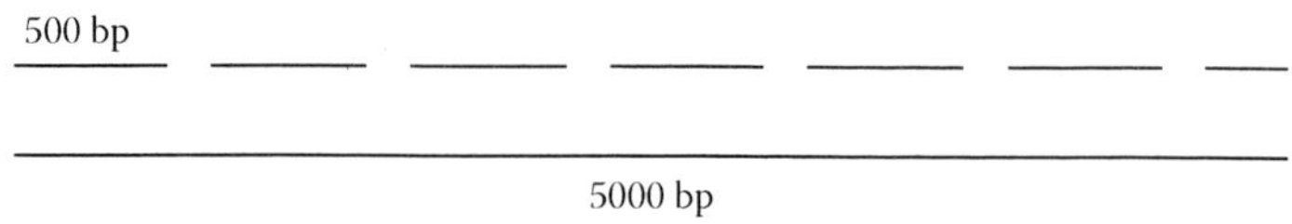

Thus, for 0.5 μg of the vector, 0.5 μg of the insert will automatically mean 10 times more insert than vector, hence a 1:10 vector:insert ratio.

D.2 Chapter 3 Answers

D1. A genomic library has to be screened to obtain the promoter for this gene, however, we need a probe to screen the genomic library. Commonly, the 5′ end of a coding sequence is used as a probe to screen for the promoter, however, in this question we do not have the coding DNA sequence. Therefore, the problem is a multistep problem.

If we only have the protein sequence for the gene, and not the coding one, the initial step will be to try and obtain the potential coding sequence for this gene; if possible, amplify the cDNA. To do that, we can use degenerate primers to amplify the coding sequence for the gene. The degenerate primers will be designed using the protein sequence information, as follows:

(Only the first and last five amino acids are shown here, since a 15-base primer for a forward and reverse sequence is sufficient for this sample question.)

M	T	E	Y	K	......
ATG	Acx	GA(A/G)	TA(T/C)	AA(A/G)	

K	C	I	I	M
AA(A/G)	TG(T/C)	AT(T/C/A)	AT(T/C/A)	ATG

D.3 Chapter 4 Answers

D1. (a) The simplest way to analyze phosphorylation of a protein is by a Western blot through the use of a phosphor-specific antibody for the protein in question (in this case, the anti-phospho-STAT antibody as the primary antibody). An appropriate secondary antibody (with HRP conjugate for chemiluminescence assays) will then be chosen from the list. Cells will be treated with or without growth factor, after which cell lysates will be prepared, protein extracts loaded on the SDS-PAGE gel and subjected to Western blotting.

(b) Localization can be studied either by cellular fractionation and Western blotting (too labor-intensive for a preliminary study), or by immunofluorescence. Again, cells will be treated with or without a growth factor, fixed onto microscope slides, and permeated (for the antibody to penetrate the cell), treated first with primary antibody (in this case, anti-phospho-STAT antibody as primary antibody), then an appropriate secondary antibody (with a conjugated fluorescent dye). Ideally, a DNA stain should also be added (propidium iodide, DAPI or similar, but the secondary antibody should have different absorption and emission maxima).

D.4 Chapter 5 Answers

D3. The first thing to do (this part was extensively discussed in previous chapters) will be to clone the coding sequence of Ras into a mammalian expression vector. In order to do this, either normal mammary tissue or breast tumor tissue can be used, RNA is extracted, reverse transcribed into cDNA and Ras-specific primers (designed based on the sequence information available on gene databases, NCBI) will be used to amplify the Ras cDNA. Appropriate restriction enzymes will be used to clone this sequence to a vector of, for instance, pCMV series. The next step will be to create the oncogenic mutant of Ras, through site-directed mutagenesis (the nature of the oncogenic mutation is not given in the question, therefore more detail cannot be presented).

(a) After having obtained both wild-type and oncogenic forms of the Ras on pCMV or a similar vector, the next step is to determine, as the question asks, whether the oncogenic Ras downregulates the BRIP1 expression in breast tumor cells. The simplest thing to do is to transfect the breast tumor cell lines with either an empty control vector (pCMV only), wild-type Ras, or oncogenic Ras. Afterward, the RNA will be isolated from all the samples and converted into cDNA using reverse transcriptase, followed by BRIP1-specific PCR, that is, RT-PCR for BRIP1. A housekeeping gene such as GAPDH or actin should also be amplified as a control.

(b) In order to do this, first BRCA1 and BRIP1 should be cloned into an expression vector (if Ras is tagged, then BRCA1 and BRIP1 should each be cloned into a different tagged vector). We want to know whether BRCA1 can still bind to chromatin in the absence of BRIP1; and both of them are already cloned in different vectors (say, BRCA1-Flag and BRIP1-Myc). Breast tumor cell lines will be transfected with either BRCA1 alone, or BRCA1 and BRIP1 together. For the chromatin immunoprecipitation assays, the protein extracts from each cell (already fixed) will be subjected to immunoprecipitation with an anti-Flag antibody, and following appropriate treatments (such as reverse cross-linking and protein degradation), the remaining DNA will be used to amplify the target sites for BRCA1 (not given in the question, therefore we can assume that we already have primer pairs specific for each target sequence). The PCR products will then be run on a DNA agarose gel and analyzed. (We would expect to see amplification when both BRCA1 and BRIP1 are co-transfected to the cells, but not in other samples.)

D.5 Chapter 6 Answers

D1. The two coding sequences will be cloned into two different fluorescent expression vectors; for example, JGF1 to pEGFP, and JGF2 to pDsRed. The cloning will be done similarly for both: the JGF1 and JGF2 coding sequences will be obtained by RT-PCR amplification from cDNA reaction obtained from target cells (cell type not mentioned here). The primers used for amplification of either growth factor should contain restriction enzyme motifs that match the MCS of the relevant vector (you should check that these motifs are not present within the coding sequence). The vectors will also be digested with the corresponding restriction enzymes, followed by alkaline phosphatase digestion and purification. The cDNAs and vectors will be ligated and transformed into bacteria. Resulting colonies will be screened for positives (typically using the same restriction enzymes used for cloning).

GFP-JGF1 and RFP-JGF2 plasmids thus prepared will be co-transfected to the target cell (cell type not mentioned in the question), and FRET analysis will be performed in live cells, by exciting the GFP (using wavelength that gives maximal absorption for GFP), and by detecting fluorescence from RFP (detecting emitted wavelength of maximal emission for RFP).

D.6 Chapter 8 Answers

D2. First, let us emphasize that from Chapter 7 onward, the questions are far more hypothetical than before, and may have more than one possible answer strategy. In this question, since no data are given, it is only appropriate to try and provide a theoretical approach or strategy. In this hypothetical biotech scenario, your company wants to "produce" liver from ES cells, however, differentiated liver cells should be labeled with GFP (presumably so as to separate them from other cell types or undifferentiated cells). Therefore, the simplest solution could be stable genetic manipulation of the ES cells in such a way that a GFP coding sequence driven by a liver cell-specific promoter is integrated (*knocked in*) to the ES genome, using the methods described in the chapter.

D.7 Chapter 9 Answers

D1. As with previous chapters, there is no single method or strategy to solve these questions. There may be a multitude of different approaches. Question 1 is "solved" here using one possible approach. In this question, a tobacco enzyme has to be expressed in transgenic rice for large-scale production. The coding sequence for this tobacco enzyme XYZ is given (the hypothetical *gene* sequences are always kept short due to space constraints, since these are exam questions). However, it is not sufficient to just express this enzyme, it needs to be also purified, and then expression of the enzyme needs to be confirmed, for which a tag is required (for affinity purification and Western blot for monitoring expression, for example).

Any plant modification strategy, for manipulation as well as choice of vector, described in this chapter could be used. pCAMBIA vector shown in this chapter (Figure 9.2) contains a His tag for purification, so the easiest solution could be the cloning of the coding sequence for XYZ enzyme in place of the reporter gene, avoiding any frameshifts so that His-nickel affinity could be used for purification. This vector can then be delivered to the plant using a gene gun, for example, and keeping the plants under selection pressure. Anti-His antibodies will also be used to confirm the expression of the His-tagged XYZ enzyme in rice prior to purification.

Index

D

E

F

T

W

X

Y

Z